PHYSICAL CHEMISTRY
(STATISTICAL MATHEMATICS)

PHYSICAL CHEMISTRY
(STATISTICAL MATHEMATICS)

By

D.K. Jha

DISCOVERY PUBLISHING HOUSE PVT. LTD.
NEW DELHI-110 002

First Published - 2009

Reprinted - 2016

ISBN: 978-81-8356-445-8

Physical Chemistry
(Statistical Mathematics)

Published by:

DISCOVERY PUBLISHING HOUSE PVT. LTD.
4383/4B, Ansari Road, Darya Ganj
New Delhi-110 002 (India)
Phone: +91-11-23279245, 43596064-65
Fax: +91-11-23253475
E-mail: discoverypublishinghouse@gmail.com
sales@discoverypublishinggroup.com
web: www.discoverypublishinggroup.com

Printed at:
Infinity Imaging Systems
Delhi

Contents

Preface

Physical Chemistry is concerned with the fundamental features of chemistry, as spelled out by its basic physical laws. It permeates all of the modern chemistry, along with many adjoining areas. It also provides the basis of modern methods of measurements, within all areas of chemistry. It is the application of physics to microscopic, atomic, subatomic and particulate phenomena in chemical systems, within the field of chemistry. It is mostly defined as a large field of chemistry, in which several subconcepts are applied. Physical Chemistry is mostly referred to as a macromolecular doctrine, as many a concept, on which physical chemistry was founded are composed entirely of macromolecular concepts, like colloids.

Physical chemistry is concerned with the structure and dynamics of atoms and molecules, and in particular, with the development of quantitative descriptions of processes, which occur at the submicroscopic level.

Realising the need of a comprehensive book on 'Physical Chemistry', which provides thorough knowledge on the subject, to the concerned readers, we have made this humble effort. Hopefully, it would prove to be beneficial for the students and general readers. Positive suggestions from the readers are heartily welcome.

Significance of Physical Chemistry

In Chemistry, the primary school curriculum has to a considerable extent become integrated and a large number of good primary school teachers possess a broad background, which enables them to guide their pupils' learning on a variety of topics as often based as the surroundings of the school. However secondary school curriculum generally consists of a number of separate subjects having little or no coordination between them. This may largely be due to the training received by secondary school teachers and to the public examination system which a strongly subject bounded. An attempt has been made in recent years to bring about an integrated curriculum which has helped to bring various science subjects closer but no effort has been made to consider other areas such as languages, mathematics and social sciences.

In fact, Chemistry has a significant place in the discipline of science. In many countries of the world, the primary school curriculum bears little relation to that of fifty years or so ago. Then the subjects were reading, writing and arithmetic. Now the curriculum is achieved much more as a whole.

Scientific Relationship

Mathematics and chemistry are closely related to each other. Actually speaking mathematics is considered as the mother of all sciences. A thorough knowledge of some fundamentals of mathematics is very useful in understanding certain concepts of chemistry. A closer coordination between the chemistry teacher and mathematics teacher makes the job of teaching chemistry easier. In physical chemistry such topics as thermodynamics, chemical kinetics, radioactivity, etc. Can only be properly understood by using certain mathematical equations. For derivation of such equations the students must be familiar with various sign used for representing certain mathematical operations. Thus we conclude that there is a close relationship and so there is a correlation between chemistry and mathematics.

Science and Social Sciences

Chemistry is a highly useful subject for the present day society. Many an inventions in chemistry have a lot of social implications and influences the social thinking of individuals. Knowledge of chemistry is quite useful in dispelling superstitions. The contribution of chemistry in development of society is visible in all walks of our life. Many a luxuries which have now become essential for comfortable living owe their origin to knowledge of chemistry.

Teacher can refer to such contributions of chemistry while teaching the social sciences. In history reference can be made to various inventions in chemistry which were used to fight or win wars. Geography depend highly on chemistry for some of its aspects. The two subjects geography and chemistry overlap in various areas particularly in areas of study of rocks, atmosphere, hydrosphere, lithosphere, minerals, rain etc. Present day geography is considered as one of the science subjects.

Collaboration with Physics

Chemistry and physics both are branches of science and they have a large number of common concepts. Many a laws of chemistry can be quite useful for explanation of certain important concepts in physics. The illustration of common topics in chemistry and physics is given by topics such as nuclear physics, thermal physics, atomic physics, etc. Many a methods of chemistry are used for carrying out the experiments in physics. This points to a scope of great cooperation between chemistry and physics teachers.

Collaboration with Biology

The correlation between chemistry and biology is so large that at present we came across such subjects a "biochemistry". There are many a topics in biology which are quite dependent on knowledge of chemistry, *e.g.* biomolecules, working of various human systems such as blood circulation, digestive system etc. The knowledge of chemistry is helpful in understanding various diseases and in helping to cure/prevent such diseases.

From the above we find a lot of correlation between chemistry and biology. For a better teaching there should be close cooperation between the chemistry teacher and biology teacher.

Significance of Creativity

Chemistry like physics is an experimental science and so it has grown through inventions and discoveries. These require a lot of creativity. It is possible to fulfil the creative urge of students if chemistry is taught to them using the method 'learning by doing'. For this the teacher is expected to impart chemistry instructions in such a way that students are actively involved in all activities and it places a good deal of responsibility on chemistry teachers.

Creativity has been defined in various ways, however there is one thing in common in all these, *i.e.* creativity is a process of change, of getting away from main track, of sensing gaps or disturbing missing elements.

A definite correlation has been shown between creativity and intelligence. A high intelligence does not mean high creativity. According to Guilford, creativity represents patterns of primary abilities, patterns which can vary with different spheres of creative ability. It is generally believed that creativity consists of about 120 abilities, the most important of these being sensitivity to problems, fluency of ideas, originality and redefining. Some of personality characteristics found in creative persons are:

1. Curiosity
2. Ambition
3. Drive
4. Independence of Judgement
5. Self assertion
6. Imagination
7. Initiative
8. Concern for basic problems
9. Openness
10. High ego strength
11. Emotional stability
12. Less talkative
13. Abstract thinking
14. Non-conformist attitude
15. Capability to take risk.

It is possible to foster creativity, through chemistry, in children if we make use of scientific methods in teaching of chemistry. The important of such methods are problem solving method, project method, laboratory method, etc. We can say that most chemists were creative because they relied on method of discovering new knowledge. For fostering creativity in children the chemistry teacher is expected the perform varied roles. He is expected to perform the following roles:

1. He should give due consideration to questions and ideas of his students.
2. He should put provoking questions in class.
3. He must be able to recognise originality and should value such an originality.
4. He must foster in his students an ability to elaborate a given point.
5. He should set more questions and problems for experimentation.
6. He should help in developing creative ideas.
7. He should have guided and planned experiences.
8. He should emphasise for research of truth through experimental research.
9. He should choose same investigatory projects in chemistry which give his children an opportunity of self-direction.
10. He should encourage his students to improvise chemistry apparatus and experiments.

Various Systems

For improving the quality of science learning in non-formal system of education, Vikram Sarabhai Community Centre was established in Ahmedabad in 1963. The Centre is one of the pioneer organisations in the country providing a variety of out-of-school activities in science for students, teachers and community. It has a team of highly skilled staff which acts as a nucleus and catalyst for various programmes undertaken by the Centre.

The Centre conducts research and innovative programmes for improving education and community life. These programmes include studies on science and mathematics, environmental studies, integrated science and science learning improvement programmes through enquiry approach, mathematics laboratory, teacher orientation, designing and development of teaching and learning material packages.

The Centre also organises programmes for rural as well as urban community. These programmes are related to the problems of pollution, health, security, population, communication, settlement and values. The Centre is basically a community centre where people come with their children and learn science where interested teachers and scientists experiment new ideas in teaching and learning.

The Centre organises science seminars, film video shows, popular lectures, exhibitions, sky-gazing through a telescope, etc. The Centre has a library, laboratories, science museum, workshop, science playground, mass media and A.V. facilities for the community. The Centre provides facilities in rocketry and electronics hobbies to children. In science playground, the children get a glimpse of science through play toys, colour filter towards musical pipes, sand, pits, water pond and evolution pillar.

The centre has also started some small extension centres to the rural areas. A mobile van equipped with a laboratory and A.V. materials tours different villages. Science club activities are organised in rural areas based on emphasis on the environmental awareness.

The Kerala Shastra Sahitya Parishad (KSSP) is a voluntary organisation. It was established in 1963. KSSP has around 10,000 members comprising scientists, doctors, engineers, social scientists, teachers, students, workers, peasants and technicians. It has 600 units all over Kerala.

Objectives

1. To popularise science amongst society.
2. To generate science literacy amongst people.

3. To increase community involvement for developing scientific temper in the society.
4. To develop rural technology in the field of energy.
5. To organise health camps, classes and audio-visual campaigns on a wide scale.

Activities: The society has about eight major areas of activities: (a) Publications, (b) Non-formal Education, (c) Formal Education, (d) Environmental Bridge, (e) Research and Development Wing, (f) Rural Science Forums, (g) Health Brigade, (h) Art and Science.

The details of these activities are as follows.

Publications: KSSP prints a variety of scientific periodicals and books meant for popularisation of science and generation of science literacy amongst the people.

These include:

1. Eureka—monthly magazine for primary classes.
2. Sastrakeralam—monthly magazine for secondary School children.
3. Sastragathy—monthly magazine for adults.
4. Parishad Vartha—monthly bulletin for members.

Non-formal Education: These activities cover a wide spectrum, the main ones being 'Science Campaign' and 'Science Centre.'

Formal Education: KSSP promotes a number of activities aimed at improving science clubs, talent tests and promotion of awareness about the education system amongst the public. The talent tests are:

1. Eureka Talent Tests—at elementary level.
2. Sastrakeralam Quiz—at high school level.
3. Sastragathi Talent Tests—at college level.

Environmental Bridge: KSSP was involved m Silent Valley Campaign, Social Forestry Programmes and campaigns against industrial pollution.

Research and Development Wing: Its responsibility is to develop appropriate rural technology in the field of energy, environment, etc. A high efficiency *Chulha* (Stove), developed by them has been widely propagated.

Rural Science Forums: KSSP has initiated these forums to prompt villagers to think on their own about their problems and solutions.

Health Brigade: KSSP organises health camps, classes and audio-visual campaigns on a wide scale.

Art and Science: It organises Sastra Kala Gatha and Bharat Kala Gatha.

Ekalavya Science Teaching Project (ESTP) started in 1972 in sixteen rural middle schools of district Hoshangabad in Madhya Pradesh for teaching science through environment based discovery approach. This project was started by Kishore Bharati, in collaboration with Friends Rural Centre, Rasulia with the support of the Department of Education, Government of Madhya Pradesh. A large number of teachers and scientists from various institutions and organisations such as the All India Science Teachers Association, Physics Study Group; Mumbai (Bombay) Municipal Corporation; Gandhi Vidyapeeth, Vedehi, Surat District; Lok Bharti, in Gujarat; The Space Application Centre, Ahmedabad; Universities of Delhi, Rajasthan and Indore; The Tata Institute of Fundamental Research, Mumbai (Bombay); Indian Institute of Technology, Kanpur; NCERT, DAV College of Education, Abohar (Punjab) etc. participated in the development of curriculum, workbooks, science kit, other materials and training of teachers.

In 1978 this programme of science teaching was extended to all the 206 middle schools of District Hoshangabad.

Objectives

1. Implementation of introducing innovations as envisaged in Ekalavya Project within the given framework of the Government school system.
2. Encouraging science teaching through discovery approach in Indian schools.
3. Providing science education experiences through environment.
4. Developing ability among students for applying scientific methods in different situations.
5. Developing scientific attitude (scientific temper) among the students.

Curriculum: Keeping in view the objectives of this Ekalavya experiences the curriculum of science teaching has been on process approach rather than product approach. The process approach of learning science provides numerous opportunities to children to explore scientific phenomena of their local environment. Most of the curricular contents have been taken from their environment.

Advanced scientific concepts, such as abstract chemical symbols, theoretical concepts of atomic and molecular structure, and human anatomy, etc. have not been included in the curriculum because these concepts are beyond the students direct interaction with the environment.

The selection of curricular content is dependent upon: (a) relatedness to environment, (b) relatedness to the needs, interests and mental level of the students, and (c) possibility of the application of discovery approach.

Through the above-mentioned procedure the curricular contents for classes sixth, seventh and eighth was developed. Some of the examples of the curricular contents for various classes are given below:

Class VI	Kuchh Khel Khilwar
	Samuh Banana Sikho
	Hamari Phaslen aur Samuhikaran
	Vidyut
	Ganak Ke Khel
	Jeev Jagad me Vividhata
	Mini, Pathar aur Chattane
Class VII	Ek Majedar Khel
	Jar aur Patti
	Keeron Ki Duniya
	Phaslon Ke Dushman
	Apni Haddi Pahachano
	Aakash Ki Or
	Taraju Ka Sidhant
Class VIII	Jantuon Ka Jivan Chakkar
	Phool aur Phal
	Paudho me Prajanan
	Vargikaran Ke Niyam
	Jantuon ka Vargikaran
	Gasen

Work Book and Science Kit Materials: In this programme, the workbook is introduced in place of textbooks, which is process based. Principles of science are discovered through experiments. Science Kit is very conducive for discovery approach to science teaching.

Teaching Method: Discovery approach of science teaching is the main teaching method followed in this programme. Students learn science through enquiry approach especially by experimentation, discussion and field trips.

The whole class is divided into subgroups of four students each known as a Toli. This Toli pattern is also followed in their teacher training programmes. The students perform experiments in their respective tolies, collect and analyse data and draw conclusions on the basis of the guidelines given in the workbook.

Examination: In this Eklavaya Experience the examination is not based on rote memory or recall, etc. Independent observation, data collection, data analysis and drawing conclusions have been given due weightage. It also seeks to test the extent of a pupil's readiness to innovate through physical experimentation.

The examination is conducted to test three basic elements of science teaching, namely, scientific skills, scientific attitude (scientific temper) and understanding of scientific concepts and principles.

Theory and Practice Collaboration

There is a lot of correlation between chemistry and some of the work experience subjects. It is due to this that in many schools chemistry-teachers are assigned the duties which require them to take some work experience subjects. A few such subjects are:

1. Candle making.
2. Itching.
3. Engraving.
4. Chalk making.
5. Preparation of shoe polish and nail polish.
6. Preparation of soaps and detergents.
7. Preparation of antiseptics, cosmetics, etc.

Importance of Collaboration

No subject can be taught in isolation and so is the case with teaching of chemistry. For an effective learning full advantage must be taken of various correlations and applications of chemistry. In addition to correlation of chemistry with other school subjects and daily life, a lot of correlation is possible with other science subjects. Artificial division of science into various branches is a matter of convenience and not of necessity. Based upon this premise, many educators advocate the implementation of curricula based upon the correlation between various subjects. These kinds of curriculum give more meaning to our class room instructions. Various inventions in chemistry have contributed a lot to the social and physical advancement of our society. Chemistry has contributed a lot to development of some other subjects.

Language Collaboration

Chemistry in closely related to language in which it is taught. This correlation arises because of the fact that language provides not simply a way of communicating with others but it is also the vehicle of thought. A student can not grapple with a scientific problem without the use of words. This brings about the importance of encouraging the student to master a language both in its spoken and written forms.

Practical work in science provides a very good opportunity for development of a language. It can be developed by discussion between the student and the teacher and also by discussion amongst students themselves. The written form of the language can be developed by encouraging the students to make their own record of practical work, may be in the form of a diary, instead of copying it from the book or blackboard.

The Right Context

Science education forms an integral part of our school curriculum upto the secondary level in the process of Universalisation of Science Education. The National Policy on Education (NPE)-1986 has laid considerable emphasis on strengthening the science education in the school education system. The current situation of teaching of school science content is as Environmental Studies at the lower primary level (classes I—V). Integrated approach of teaching science is followed at the upper primary stage (classes VI—VIII) and secondary stage (classes IX—X) by integrating all disciplines of Science in a natural fashion. The NPE-86 envisages extension through every effort of Science Education not only to those who are in the formal school system but also to those who have remained outside the system under non-formal and adult education programmes.

The qualitative improvement in science education depends on many vital components. The teacher is considered as a crucial factor in the teaching-learning process, developing positive attitudes in the learners for better achievement and the formulation and implementation of science education programmes. The teachers have to discard their traditional methods and usual practices in relying entirely on textbooks. The teaching has to be integrated with environment based on real life situations using local experiences, expertise and resources. The classroom territory has to be expanded over the whole environment so that the activities become supplementary to classroom teaching.

If such an approach is systematically implemented with mobilisation of needed resources, it is very much likely that there may be an improvement in our science education at the school stage. Several attempts are being made to improve science teaching in our country, both at the formal as well as non-formal sectors of school education.

Some of the innovative experiences in science education at the school level include Nehru Science Exhibition, Science Museums and Mobile Science Vans, Vikram Sarabhai Science Centre Ahmedabad, Kerala Shastra Sahitya Parishad and Ekalavya Project, etc. In such innovative experiences and science activities, the learning is fitted to the abilities and interests of learners as there exists an opportunity for individuals initiative, independent or collective study and creativity. A large number of such activities are organised in India. A few of the major activities are listed and described briefly here.

Profit of Exposure

November 14 is the birth anniversary of Pandit Jawahar Lal Nehru, our first Prime Minister. Children call him Chacha Nehru, as he used to love them much. We celebrate Children's Day on November 14. On this occasion Science Exhibition is also organised by the NCERT. This exhibition is Science Exhibition. To start with, this exhibition used to be organised at New Delhi where Pandit Nehru used to live, when he was our Prime Minister. Now this exhibition is organised in different states. Children from country, from all States and UTs participate in this exhibition. Nehru Exhibition is organised at the national level.

Every year NCERT announces a main theme and sub-themes for Nehru Exhibition. Children make their exhibits on these sub-themes.

Main Theme: Science in our Environment

1. Agriculture, Horticulture, Farming and Animal Husbandry.
2. Conservation of the Environment
3. Health
4. Energy Conservation and Needs
5. Astronomy
6. Town and Village Planning
7. Machines in the Service of Rural Areas
8. Teaching Aids for Science and Mathematics
9. Innovations.

This information along with the last date of submitting the entry with the dates of exhibition reaches every school of the country, from States and UTs and from States and UTs to all schools. On information children start working on their projects—static or work in investigatory science projects under the guidance and supervision of the teachers. If a school has a Science Club, it becomes quite active after information under the supervision and guidance of the Science Club.

Responsibility by Instructions

Very often in Nehru Science Exhibition we see some static work or some noble experiments demonstrated by the students with lines and graphs. These are projects but not Investigatory Science Projects. Students should be encouraged and motivated to work on some of science projects, and bring them to Nehru Science Exhibition.

A project may be any purposeful activity. It may be a model working, or experiment.

A project which involves investigation, discovery and finding out which was not known to the student before, is an investigatory investigation is much more than the repetition of a standard Experienced student is to decide what experiments are necessary. He may have to design his own apparatus, if that is not available in the laboratory. He has to search for the appropriate principles, laws, formulae, apparatus and data, and originate a solution to a problem. The student has to behave like a scientist.

Working on an investigatory science project is the way a student can learn science by project method which involves some steps of scientific method like Problem, Hypotheses and Experiment.

Every year more and more investigatory science projects are being seen in Nehru Science Exhibition. It is very encouraging. You should also encourage your students to work on investigatory science projects under your guidance and send it to Nehru Science Exhibition through District and State Science Exhibitions and Fairs.

Profits from Museums

You must have seen science museums. They are very effective and interesting sources of learning science. What experiences do you get from a science museum?

Objectives of Science Museum: Main objectives in establishing science museums are:

1. To help young science learners in understanding concepts of science by play way method.
2. To provide a glimpse of past as well as an insight into the future.
3. To help schools in their class activities by providing them with a number of equipments and specimens which are otherwise difficult for a single school to procure.
4. To arrange extension activities such as field trips, lectures, film shows and exhibitions for the students as well as public.

During the last decade or so, a couple of science museums have been set up in the country including one at Delhi, the Natural History Museum. In Delhi's Pragati Maidan, National Science Centre (National Council of Science Museums) has also been set up few years back. It has several units. 'FUN GAMES' and 'ENERGY' units are very interesting for students. Delhi also has two Primary Science Museums:

1. Municipal Corporation Children Resource Centre (MCCRC) at R.K. Puram, Sector VI.
2. New Delhi Municipal Council (NDMC) Science Centre at Lakshmibai Nagar.

There are good science museums in various parts of the country—Mumbai (Bombay), Bangalore, Kolkata (Calcutta). All these science museums are doing very good Job, carrying out various innovative activities, in science for the improvement of science education. Let us discuss some science museums a little bit in detail.

Nehru Science Centre: The Nehru Science Centre (NSC) is established in Mumbai (Bombay) by the National Council of Science Museums.

The most important and attractive part of the NSC is a 'Science Park' for children. With green surroundings, the Children's Science Park has exhibited on time, motion, energy, power and work. Also, there are models of railway engines, tram cars, aeroplanes, steam lorries, a windmill and a sun dial. There are birds, animals and fish to acquaint children with nature. While children enjoy the Science Park the most, it also helps them to understand 'what' 'why' and 'how' of the queries, questions and problems haunting their minds.

Nehru Science Centre, Mumbai (Bombay) is basically multi-disciplinary in character. Collection of antique exhibits of historic value, presentation of the same through permanent and temporary exhibitions on selected themes, extension activities offering multiple avenues of learning, enjoyment and training to the student community as well as the public, taking science to rural areas through mobile science vans, aiming towards interacting mode of presentation of themes are some of the ways in which the NSC operates. The Science Centre also offers a gallery on 'light and sight'. It presents different principles involved in file process of 'seeing and the vision.' A survey is made of vision, its defects, its importance, its complexities and varieties. Nehru Science Centre also organises extension activities such as science extension in rural areas, film video shows, science seminars for schools, films, video cassettes loan service, amateur weather station, amateur radio classes for children, sky observation programme, astronomical camps, popular science lectures, special science film festivals, aeronautic modelling programmes and training camps for under-privileged children.

Visweshwaraya Industrial and Technological Museums: The Visweshwaraya Industrial and Technological Museum is established in Mumbai (Bombay). It organises various activities and programmes such as motive power gallery (science museum), teacher training, hobby centre, student's science seminars, science quiz, science fair, temporary science exhibitions, science demonstration lectures, mobile science exhibitions, film shows, popular science lectures, etc. The museum has also a regional science centre at Gulbarga.

If you have a science museum at the place where you teach, plan a visit to that museum. If your students go to some places, where there are science museums, ask them to visit them with their parents or if you arrange a field trip to any place, where there is a science museum, take your students there, and see how much science they learn—the science which is not there even in their science books.

Mobile Science Vans: Some science museums have mobile science units, museums on wheel. They are usually sent to the places, where there are no science museums. They are not as big as a science museum. They do not have as many science exhibits as a science museum has. But even this is a very effective source of learning science for students at upper primary level.

Natural History Museum, New Delhi, National Science Centre, New Delhi and Nehru Science Centre, Mumbai (Bombay) have Mobile Science Vans. You try to find out whether there is some science museum, not very far from the place you are teaching, and whether that science museum has the facility of mobile science vans. You can ask such museums to send that unit to your school. If you are successful to bring it, your students may visit the science museum even in their school, and can learn a lot of science.

Applied Statistics

Traditionally, statistics was concerned with drawing inferences using a semi-standardised methodology that was required learning in most sciences. This has changed with use of statistics in non-inferential contexts. What was considered to be a dry subject, taken only as a requirement for degrees in many fields, is now viewed enthusiastically. What was derided by some mathematical purists is now considered essential methodology in some areas:

- Scatter plots of data generated by a distribution function may be transformed with familiar tools used in statistics to reveal underlying patterns, which may lead to hypotheses in number theory.
- Methods of statistics including predictive methods in forecasting, are combined with chaos theory and fractal geometry to create video works considered to be of beauty. The process art of Jackson Pollock relied on artistic experiments whereby underlying distributions in nature were artistically revealed. With the advent of computers, methods of statistics were applied to

formalise such distribution driven natural processes, in order to make and analyse moving video art.

- Methods of statistics may be used predicatively, not inferentially in performance art, as in a card trick based on a markov process that only works some of the time, predicted using statistical methodology.
- Statistics is used to predicatively create art, for example in applications of statistical mechanics with the statistical or Stochastic music invented by Iannis Xenakis, where the music is performance specific, and does not always come out as expected, but does within a range predicted using statistics.

Mathematical Statistics

Mathematical statistics is the study of statistics from a purely mathematical standpoint, using probability theory as well as other branches of mathematics such as linear algebra and analysis.

Mathematical statistics deals with gaining information from data. In practice, data often contain some randomness or uncertainty. Statistics handles such data using methods of probability theory.

Statistics is divided into:

- *Descriptive Statistics:* The part of mathematical statistics that describes data, i.e. summarises the data and their typical properties.
- *Inferential Statistics:* The part of mathematical statistics that draws conclusions from data, i.e. checks whether the data fulfil some condition and gives guarantees on the involved uncertainty.

Mathematical statistics is the theoretical basis for many practices in applied statistics.

Exact Statistics

Exact statistics, such as that described in exact test, is a branch of statistics that was developed to provide more accurate results pertaining to statistical testing and interval estimation by eliminating procedures based on asymptotic and approximate statistical methods. The main characteristic of exact methods is that statistical tests and confidence intervals are based on exact probability statements that are valid for any sample size.

When the sample size is small, asymptotic results given by some traditional methods may not be valid. In such situations, the asymptotic p-values may differ substantially from the exact p-values. Hence, asymptotic and other approximate results may lead to unreliable and misleading conclusions.

Exact statistical methods help avoid some of the unreasonable assumptions of traditional statistical methods, such as the assumption of equal variances in classical ANOVA. They also allow exact inference on variance components of Mixed Models.

When exact p-values and confidence intervals are computed under a certain distribution, such as the normal distribution, then the underlying methods are referred to as Exact Parametric Methods. The exact methods that do not make any distributional assumptions are referred to as Exact Non-parametric Methods. The latter has the advantage of making lesser assumptions whereas, the former tend to yield more powerful tests when the distributional assumption is reasonable. For advanced methods such as higher-way ANOVA, Regression, and Mixed Models, only exact parametric methods are available.

Applied Mathematics

Applied mathematics is a branch of mathematics that concerns itself with the mathematical techniques typically used in the application of mathematical knowledge to other domains.

Divisions

There is no consensus of what the various branches of applied mathematics are. Such categorisations are made difficult by the way mathematics and science change over time, and also by the way universities organise departments, courses and degrees.

Historically, applied mathematics consisted principally of applied analysis, most notably differential equations, approximation theory (broadly construed, to include representations, asymptotic methods, variational methods, and numerical analysis), and applied probability. These areas of mathematics were intimately tied to the development of Newtonian Physics, and in fact the distinction between mathematicians and physicists was not sharply drawn before the mid-19th century. This history left a legacy as well; until the early-20th century subjects such as classical mechanics were often taught in applied mathematics departments at American universities rather than in physics departments, and fluid mechanics may still be taught in applied mathematics departments.

Today, the term *applied mathematics* is used in a broader sense. It includes the classical areas above, as well as other areas that have become increasingly important in applications. Even fields such as number theory that are part of pure mathematics are now important in applications (such as cryptology), though they are not generally considered to be part of the field of applied mathematics *per se*. Sometimes the term *applicable mathematics* is used to distinguish between the traditional field of applied mathematics and the many more areas of mathematics that are applicable to real-world problems.

Mathematicians distinguish between applied mathematics, which is concerned with mathematical methods, and applications of mathematics within science and engineering. A biologist using a population model and applying known mathematics would not be doing applied mathematics, but rather using it. However, non-mathematicians do not usually draw this distinction.

The success of modern numerical mathematical methods and software has led to the emergence of computational mathematics, computational science, and computational engineering, which use high performance computing for the simulation of phenomena and solution of problems in the sciences and engineering. These are often considered interdisciplinary programmes.

Some mathematicians think that statistics is a part of applied mathematics. Others think it is a separate discipline. Statisticians in general regard their field as separate from mathematics, and the American Statistical Association has issued a statement to that effect. Mathematical statistics provides the theorems and proofs that justify statistical procedures and it is based on probability theory, which is in turn based on measure theory.

The line between applied mathematics and specific areas of application is often blurred. Many universities teach mathematical and statistical courses outside of the respective departments, in departments and areas including business and economics, engineering, physics, psychology, biology, computer science, and mathematical physics. Sometimes this is due to these areas having their own specialised mathematical dialects. Often this is the result of efforts of those departments to gain more student credit hours and the funds that go with them.

Usefulness

Historically, mathematics was most important in the natural sciences and engineering. However, in recent years, fields outside of the physical sciences have spawned the creation of new areas of mathematics, such as game theory, which grew out of economic considerations, or neural networks, which arose out of the study of the brain in neuroscience, or bioinformatics, from the importance of analysing large data sets in biology.

The advent of the computer has created new applications, both in studying and using the new computer technology itself (computer science, which uses combinatorics, formal logic, and lattice theory), as well as using computers to study problems arising in other areas of science (computational science), and of course studying the mathematics of computation (numerical analysis). Statistics is probably the most widespread application of mathematics in the social sciences, but other areas of math are proving increasingly useful in these disciplines, especially in economics and management science.

Status in Academic Departments

Academic institutions are not consistent in the way they group and label courses, programmes, and degrees in applied mathematics. At some schools, there is a single mathematics department, whereas others have separate departments for Applied Mathematics and Pure Mathematics. It is very common for statistics departments to be separate at schools with graduate programmes, but many undergraduate-only institutions include statistics under the mathematics department.

Many applied mathematics programmes (as opposed to departments) consist of primarily cross-listed courses and jointly-appointed faculty in departments representing applications. Some Ph.D. programmes in applied mathematics require little or no coursework outside of mathematics, while others require substantial coursework in a specific area of application.

In some respects, this difference reflects the distinction between "application of mathematics" and "applied mathematics".

Some universities in the UK host departments of Applied Mathematics and Theoretical Physics, but it is now much less common to have separate departments of pure and applied mathematics. Schools with separate applied mathematics departments range from Brown University, which has a well known and large Division of Applied Mathematics that offers degrees through the doctorate, to Santa Clara University, which offers only the M.S. in applied mathematics. Research universities dividing their mathematics department into pure and applied sections include Harvard and MIT.

At some universities there is some tension between applied and pure mathematics departments. One reason is that pure mathematics is often perceived as having a higher intellectual standing. Another reason is a different level of compensation, as applied mathematicians are often paid more. Applied mathematics also enjoys better opportunities to bring external funding from many sources, not limited to the Division of Mathematical Sciences at the National Science Foundation (NSF) like much of pure mathematics. External funding is highly valued at research universities and is often a condition for faculty advancement. Similar tensions can also exist between statistics and mathematics groups and departments.

Basic Statistic Topics

Statistics is a mathematical science pertaining to the collection, analysis, interpretation and presentation of data. It is applicable to a wide variety of academic disciplines, from the physical and social sciences to the humanities; it is also used and misused for making informed decisions in all areas of business and government.

The following outline is provided as an overview of and introduction to statistics:

Basic Statistics Concepts

Analysis of Variance: In statistics, ANOVA is short for analysis of variance. Analysis of variance is a collection of statistical models, and their associated procedures, in which the observed variance is partitioned into components due to different explanatory variables.

The initial techniques of the analysis of variance were developed by the statistician and geneticist R.A. Fisher in the 1920s and 1930s, and is sometimes known as *Fisher's ANOVA* or *Fisher's analysis of variance*, due to the use of Fisher's F-distribution as part of the test of statistical significance.

Binomial Distribution: In probability theory and statistics, the binomial distribution is the discrete probability distribution of the number of successes in a sequence of n independent yes/no experiments, each of which yields success with probability p. Such a success/failure experiment is also called a *Bernoulli experiment* or *Bernoulli trial*. In fact, when $n = 1$, the binomial distribution is a Bernoulli distribution. The binomial distribution is the basis for the popular binomial test of statistical significance. A binomial distribution should not be confused with a bimodal distribution.

Chi-square Distribution: In probability theory and statistics, the chi-square distribution (also chi-squared or χ^2 distribution) is one of the most widely used theoretical probability distributions in inferential statistics, e.g., in statistical significance tests. It is useful because, under reasonable assumptions, easily calculated quantities can be proven to have distributions that approximate to the chi-square distribution if the null hypothesis is true.

The best-known situations in which the chi-square distribution are used are the common chi-square tests for goodness of fit of an observed distribution to a theoretical one, and of the independence of two criteria of classification of qualitative data. Many other statistical tests also lead to a use of this distribution, like Friedman's analysis of variance by ranks.

Combinatorics: Combinatorics is a branch of pure mathematics concerning the study of discrete (and usually finite) objects. It is related to many other areas of mathematics, such as algebra, probability theory, ergodic theory and geometry, as well as to applied subjects in computer science and statistical physics. Aspects of combinatorics include "counting" the objects satisfying certain criteria (enumerative combinatorics), deciding when the criteria can be met, and constructing and analysing objects meeting the criteria (as in combinatorial designs and matroid theory), finding largest, smallest, or optimal objects (extremal combinatorics and combinatorial optimisation), and finding algebraic structures these objects may have (algebraic combinatorics).

Combinatorics is as much about problem solving as theory building, though it has developed powerful theoretical methods, especially since the later twentieth century. One of the oldest and most accessible parts of combinatorics is graph theory, which also has numerous natural connections to other areas.

There are many combinatorial patterns and theorems related to the structure of combinatoric sets. These often focus on a partition or ordered partition of a set.

An example of a simple combinatorial question is the following: What is the number of possible orderings of a deck of 52 distinct playing cards? The answer is 52. (52 factorial), which is equal to about 8.0658×10^{67}.

Combinatorics is used frequently in computer science to obtain estimates on the number of elements of certain sets. A mathematician who studies combinatorics is often referred to as a *combinatorialist* or *combinatorist*.

Correlation: In probability theory and statistics, correlation, (often measured as a correlation coefficient), indicates the strength and direction of a linear relationship between two random variables. In general statistical usage, correlation or co-relation refers to the departure of two variables from independence. In this broad sense there are several coefficients, measuring the degree of correlation, adapted to the nature of data.

A number of different coefficients are used for different situations. The best known is the Pearson product-moment correlation coefficient, which is obtained by dividing the covariance of the two variables by the product of their standard deviations. Despite its name, it was first introduced by Francis Galton.

Data: Data refers to a collection of organised information, usually the result of experience, observation or experiment, other information within a computer system, or a set of premises. This may consist of numbers, words, or images, particularly as measurements or observations of a set of variables.

Degrees of Freedom: In statistics, the phrase *degrees of freedom* is used to describe the number of values in the final calculation of a statistic that are free to vary.

Estimates of statistical parameters can be based upon different amounts of information or data. The number of independent pieces of information that go into the estimate of a parameter is called the *degrees of freedom*. In general, the degrees of freedom of an estimate is equal to the number of independent scores that go into the estimate minus the number of parameters estimated as intermediate steps in the estimation of the parameter itself.

Mathematically, degrees of freedom is the dimension of the domain of a random vector, or essentially the number of 'free' components: how many components need to be known before the vector is fully determined?

The term is most often used in the context of linear models (linear regression, Analysis of Variance), where certain random vectors are constrained to lie in linear subspaces, and the degrees of freedom is the dimension of the subspace. The degrees of freedom are also commonly associated with the squared lengths (or Sum of Squares) of such vectors, and the parameters of chi-squared and other distributions that arise in associated statistical testing problems.

While introductory texts may introduce degrees of freedom as distribution parameters or through hypothesis testing, it is the underlying geometry that defines degrees of freedom, and is critical to a proper understanding of the concept. Walker (1940) has stated this succinctly:

> "For the person who is unfamiliar with *N*-dimensional geometry or who knows the contributions to modern sampling theory only from second-hand sources such as textbooks, this concept often seems almost mystical, with no practical meaning."

Descriptive Statistics: Descriptive Statistics are used to describe the basic features of the data gathered from an experimental study in various ways. They provide simple summaries about the sample and the measures. Together with simple graphics analysis, they form the basis of virtually every quantitative analysis of data. It is necessary to be familiar with primary methods of describing data in order to understand phenomena and make intelligent decisions. Various techniques that are commonly used are classified as:

- Graphical displays of the data in which graphs summarise the data or facilitate comparisons.
- Tabular description in which tables of numbers summarise the data.
- Summary statistics (single numbers) which summarise the data.

In general, statistical data can be briefly described as a list of subjects or units and the data associated with each of them. Although most research uses many data types for each unit, this introduction treats only the simplest case.

There may be two objectives for formulating a summary statistic:

1. To choose a statistic that shows how different units seem similar. Statistical textbooks call one solution to this objective, a measure of central tendency.
2. To choose another statistic that shows how they differ. This kind of statistic is often called a *measure of statistical variability*.

When summarising a quantity like length or weight or age, it is common to answer the first question with the arithmetic mean, the median, or in case of a unimodal distribution, the mode. Sometimes, we choose specific values from the cumulative distribution function called *quantiles*.

The most common measures of variability for quantitative data are the variance; its square root, the standard deviation; the range; interquartile range; and the average absolute deviation (average deviation).

When formulating a graphical display to summarise a dataset, the same two objectives may apply. A simple example of a graphical technique is a histogram, in which the central tendency and statistical variability can both be visualised.

Steps in Descriptive Statistics

1. Collect Data.
2. Classify Data.
3. Summarise Data.
4. Present Data.
5. Proceed to inferential statistics (if there are enough data to draw a conclusion).

Errors and Residuals in Statistics: In statistics and optimisation, the concepts of statistical error and residual are easily confused with each other.

A statistical error is the amount by which an observation differs from its expected value; the latter being based on the whole population from which the statistical unit was chosen randomly. The expected value, being for instance the mean of the entire population, is typically unobservable. If the mean height in a population of 21-year-old men is 1.75 metres, and one randomly chosen man is 1.80 metres tall, then the "error" is 0.05 metres; if the randomly chosen man is 1.70 metres tall, then the "error" is 0.05 metres.

The nomenclature arose from random measurement errors in astronomy. It is as if the measurement of the man's height were an attempt to measure the population mean, so that any difference between the man's height and the mean would be a measurement error.

A residual (or fitting error), on the other hand, is an observable estimate of the unobservable statistical error. The simplest case involves a random sample of *n* men whose heights are measured. The sample mean is used as an estimate of the population mean. Then we have:

- The difference between the height of each man in the sample and the unobservable population mean is a statistical error, and
- The difference between the height of each man in the sample and the observable sample mean is a residual.

Note that the sum of the residuals within a random sample is necessarily zero, and thus the residuals are necessarily not independent. The sum of the statistical errors within a random sample need not be zero; the statistical errors are independent random variables if the individuals are chosen from the population independently.

In sum:

- Residuals are observable; statistical errors are not.
- Statistical errors are often independent of each other; residuals are not (at least in the simple situation described above, and in most others).

Design of Experiments: Design of experiments, or experimental design, is the design of all information-gathering exercises where variation is present, whether under the full control of the experimenter or not. (The latter situation is usually called an *observational study*). Often the experimenter is interested in the effect of some process or intervention (the "treatment") on some objects (the "experimental units"), which may be people. Design of experiments is thus a discipline that has very broad application across all the natural and social sciences.

Frequency Distribution: In statistics, a frequency distribution is a list of the values that a variable takes in a sample. It is usually a list, ordered by quantity, showing the number of times each value appears. For example, if 100 people rate a five-point Likert scale assessing their agreement with a statement on a scale on which 1 denotes strong agreement and 5 strong disagreement, the frequency distribution of their responses might look like:

Rank	*Degree of Agreement*	*Number*
1	Strongly agree	20
2	Agree somewhat	30
3	Not sure	20
4	Disagree somewhat	15
5	Strongly disagree	15

This simple tabulation has two drawbacks. When a variable can take continuous values instead of discrete values or when the number of possible values is too large, the table construction is cumbersome, if it is not impossible. A slightly different tabulation scheme based on the range of values is used in such cases. For example, if we consider the heights of the students in a class, the frequency table might look like below:

Height Eange	*Number of Students*	*Cumulative Number*
4.5 – 5 feet	25	25
5 – 5.5 feet	35	60
5.5 – 6 feet	20	80
6 – 6.5 feet	20	100

Histogram: In statistics, a histogram is a graphical display of tabulated frequencies, shown as bars. It shows what proportion of cases fall into each of several categories. A histogram differs from a bar chart in that it is the *area* of the bar that denotes the value, not the height, a crucial distinction when the categories are not of uniform width (Lancaster, 1974). The categories are usually specified as non-overlapping intervals of some variable. The categories (bars) must be adjacent.

The word *histogram* is derived from Greek: *histos* 'anything set upright' (as the masts of a ship, the bar of a loom, or the vertical bars of a histogram); *gramma* 'drawing, record, writing'. The histogram is one of the seven basic tools of quality control, which also include the Pareto chart, check sheet, control chart, cause-and-effect diagram, flowchart, and scatter diagram. A generalisation of the histogram is kernel smoothing techniques. This will construct a very smooth probability density function from the supplied data.

Inferential Statistics: Inferential statistics or statistical induction comprises the use of statistics to make inferences concerning some unknown aspect of a population. It is distinguished from descriptive statistics.

Two schools of inferential statistics are frequency probability and Bayesian inference.

Definition:

Statistical inference is inference about a population from a random sample drawn from it or, more generally, about a random process from its observed behaviour during a finite period of time. It includes:

- Point Estimation.
- Interval Estimation.
- Hypothesis Testing (or statistical significance testing).
- Prediction.

There are several distinct schools of thought about the justification of statistical inference. All are based on some idea of what real world phenomena can be reasonably modelled as probability:

- Frequency Probability.
- Bayesian Probability.
- Fiducial Probability.

The topics below are usually included in the area of statistical inference:

- Statistical Assumptions.
- Likelihood Principle.
- Estimating Parameters.
- Statistical Hypothesis Testing.
- Revising Opinions in Statistics.
- Planning Statistical Research.
- Summarising Statistical Data.

Mean: In statistics, mean has two related meanings:

- The arithmetic mean (and is distinguished from the geometric mean or harmonic mean).
- The expected value of a random variable, which is also called the *population mean*.

It is sometimes stated that the 'mean' means average. This is incorrect if "mean" is taken in the specific sense of "arithmetic mean" as there are different types of averages: the mean, median and mode. For instance, average house prices almost always use the median value for the average.

For a real-valued random variable *X*, the mean is the expectation of X. Note that not every probability distribution has a defined mean (or variance).

For a data set, the mean is the sum of the observations divided by the number of observations. The mean is often quoted along with the standard deviation: the mean describes the central location of the data, and the standard deviation describes the spread.

An alternative measure of dispersion is the mean deviation, equivalent to the average absolute deviation from the mean. It is less sensitive to outliers, but less mathematically tractable.

As well as statistics, means are often used in geometry and analysis; a wide range of means have been developed for these purposes, which are not much used in statistics.

Median: In probability theory and statistics, a median is described as the number separating the higher half of a sample, a population, or a probability distribution, from the lower half. The median of a finite list of numbers can be found by arranging all the observations from lowest value to highest value and picking the middle one. If there is an even number of observations, the median is not unique, so one often takes the mean of the two middle values. At most half the population have values less than the median and at most half have values greater than the median. If both groups contain less than half the population, then some of the population is exactly equal to the median. For example, if $a < b < c$, then the median of the list $\{a, b, c\}$ is b, and if $a < b < c < d$, then the median of the list $\{a, b, c, d\}$ is the mean of b and c, i.e. it is $(b + c)/2$.

Mode: In statistics, the mode is the value that occurs the most frequently in a data set or a probability distribution. In some fields, notably education, sample data are often called *scores*, and the sample mode is known as the *modal score*.

Like the statistical mean and the median, the mode is a way of capturing important information about a random variable or a population in a single quantity. The mode is in general different from mean and median, and may be very different for strongly skewed distributions.

The mode is not necessarily unique, since the same maximum frequency may be attained at different values. The most ambiguous case occurs in uniform distributions, wherein all values are equally likely.

Scientific Modelling: Scientific modelling is the process of generating abstract, conceptual, graphical and or mathematical models. Science offers a growing collection of methods, techniques and theory about all kinds of specialised scientific modelling.

Modelling is an essential and inseparable part of all scientific activity, and many scientific disciplines have their own ideas about specific types of modelling. There is little general theory about scientific modelling, offered by the philosophy of science, systems theory, and new fields like knowledge visualisation.

Monte Carlo Method: Monte Carlo methods are a class of computational algorithms that rely on repeated random sampling to compute their results. Monte Carlo methods are often used when simulating physical and mathematical systems. Because of their reliance on repeated computation and random or pseudo-random numbers, Monte Carlo methods are most suited to calculation by a computer. Monte Carlo methods tend to be used when it is infeasible or impossible to compute an exact result with a deterministic algorithm.

The term Monte Carlo method was coined in the 1940s by physicists working on nuclear weapon projects in the Los Alamos National Laboratory.

Normal Distribution: The normal distribution, also called the *Gaussian distribution,* is an important family of continuous probability distributions, applicable in many fields. Each member of the family may be defined by two parameters, *location* and *scale*: the mean (average, ì) and variance (standard deviation squared, ó^2) respectively. The standard normal distribution is the normal distribution with a mean of zero and a variance of one (the red curves in the plots to the right). Carl Friedrich Gauss became associated with this set of distributions when he analysed astronomical data using them, and defined the equation of its probability density function. It is often called the *bell curve* because the graph of its probability density resembles a bell.

The importance of the normal distribution as a model of quantitative phenomena in the natural and behavioural sciences is due in part to the central limit theorem. Many measurements, ranging from psychological to physical phenomena (in particular, thermal noise) can be approximated, to varying degrees, by the normal distribution. While the mechanisms underlying these phenomena are often unknown, the use of the normal model can be theoretically justified by assuming that many small, independent effects are additively contributing to each observation. The normal distribution is also important for its relationship to least-squares estimation, one of the simplest and oldest methods of statistical estimation.

The normal distribution also arises in many areas of statistics. For example, the sampling distribution of the sample mean is approximately normal, even if the distribution of the population from which the sample is taken is not normal. In addition, the normal distribution maximises information entropy among all distributions with known mean and variance, which makes it the natural choice of underlying distribution for data summarised in terms of sample mean and variance. The normal distribution is the most widely used family of distributions in statistics and many statistical tests are based on the assumption of normality. In probability theory, normal distributions arise as the limiting distributions of several continuous and discrete families of distributions.

Null Hypothesis: In statistics, a null hypothesis (H_0) is a plausible hypothesis (scenario) which may explain a given set of data. A null hypothesis is tested to determine whether the data provides sufficient reason to pursue some alternative hypothesis. When used, the null hypothesis is presumed sufficient to explain the data unless statistical evidence, in the form of a hypothesis test, indicates otherwise — that is, when the researcher has a certain degree of confidence, usually 95 per cent to 99 per cent, that the null hypothesis does not explain the data. It is possible for an experiment to fail to reject the null hypothesis; in this case, the null hypothesis is considered sufficient to explain the data and no alternative hypothesis needs to be devised or tested. Rejecting the null hypothesis does not show that any particular alternative hypothesis explains the data; merely that it is reasonable to devise alternative hypotheses that may explain the data better.

In scientific and medical applications, the null hypothesis plays a major role in testing the significance of differences in treatment and control groups. This use, while widespread, is criticised on a number of grounds.

The assumption at the outset of the experiment is that no difference exists between the two groups (for the variable being compared): this is the null hypothesis in this instance. Examples of other types of null hypotheses are:

- That values in samples from a given population can be modelled using a certain family of statistical distributions.

- That the variability of data in different groups is the same, although they may be centred around different values.

The term was coined by the English geneticist and statistician Ronald Fisher.

Outlier: In statistics, an outlier is an observation that is numerically distant from the rest of the data. Statistics derived from data sets that include outliers may be misleading. For example, if one is calculating the average temperature of 10 objects in a room, and most are between 20 and 25 degrees Celsius, but an oven is at 350 degrees Fahrenheit, the median of the data may be 23 but the mean temperature will be 55. In this case, the median better reflects the temperature of a randomly sampled object than the mean. Outliers may be indicative of data points that belong to a different population than the rest of the sample set.

In most larger samplings of data, some data points will be further away from the sample mean than what is deemed reasonable. This can be due to incidental systematic error or flaws in the theory that generated an assumed family of probability distributions, or it can simply be the case that some observations happen to be a long way from the centre of the data. Outlier points can therefore indicate faulty data, erroneous procedures, or areas where a certain theory might not be valid. However, a small number of outliers not due to any anomalous condition is to be expected in large samples.

Estimators capable of coping with outliers are said to be robust.

P-value: In statistical hypothesis testing, the p-value is the probability of obtaining a result at least as extreme as the one that was actually observed, given that the null hypothesis is true. The fact that p-values are based on this assumption is crucial to their correct interpretation.

More technically, a p-value of an experiment is a random variable defined over the sample space of the experiment such that its distribution under the null hypothesis is uniform on the interval [0,1]. Many p-values can be defined for the same experiment.

International Statistical Institute

The International Statistical Institute (ISI) is a professional association of statisticians which seeks to "develop and improve statistical methods and their application through the promotion of international activity and cooperation". It publishes a variety of books and journals, and holds an international conference every two years. Its permanent office is located in Voorburg (near The Hague), in The Netherlands.

It serves as an umbrella group for a number of specialised sections:

- Bernoulli Society for Mathematical Statistics and Probability (BS).
- International Association for Statistical Computing (IASC).
- International Association for Official Statistics (IAOS).
- International Association of Survey Statisticians (IASS).
- International Association for Statistics Education (IASE).
- International Society for Business and Industrial Statistics (ISBIS).
- Irving Fisher Committee on Central Bank Statistics (IFC, ISI transitional Section).
- Standing Committee of Regional and Urban Statistics (SCORUS) is a subcommittee of the IAOS.

The Institute was established in 1885, and has about 2,000 elected members from government, academia and the private sector.

Journals

ISI publishes the following journals:

- International Statistical Review.
- Statistical Theory and Method Abstracts - Zentralblatt.
- Bernoulli.
- Computational Statistics and Data Analysis.
- ISI Newsletter.
- Statistics Education Research Journal.
- Statistics Surveys.

Importance of Theory

Great Orthogonality Theorem Explained:

The great orthogonality theorem can be stated as

$$\sum_{R}\left[{}_{i}(R)_{mn}\right]\left[{}_{j}(R)_{m'n'}\right]^{*} = \frac{h}{\sqrt{l_i l_j}}\,\delta_{ij}\,\delta_{mn'}\,\delta_{nn'} \qquad \text{... (1)}$$

where 'h' is the order of the group l_i is the dimension of the ith representation l_j is the dimension of the jth representation R denotes the various operations in the group, $\Gamma_i(R)_{mn}$ denotes ith irreducible representation of an element in the mth row and the nth column and $[\Gamma_j(R)_{m'n'}]^*$ denotes the complex conjugate of the factor on the left hand side. From the above equation, it is clear that in the set of matrices constituting any one irreducible representation, one from each matrix behaves as the components of a vector in h-dimensional space in such a way that all these vectors are mutually orthogonal and each is normalised so that the square of its length equals $\frac{h}{l_i}$. For example equation (1) can be split into three equations. Since the equations are split, the complex conjugate can be omitted.

$$\sum_R \Gamma_i(R)_{mn} \Gamma_j(R)_{mn} = 0, \quad \text{if } i \neq j \qquad \text{... (2)}$$

$$\sum_R \Gamma_i(R)_{mn} \Gamma_j(R)_{m'n'} = 0, \quad \text{if } m \neq m' \text{ and/or } n = n' \qquad \text{... (3)}$$

$$\sum_R \Gamma_i(R)_{mn} \Gamma_j(R)_{mn} = \frac{h}{l_i} \qquad \text{... (4)}$$

Equation (2) indicates that the vectors differ only in the fact that they are chosen from matrices of different representations and are orthogonal.

Equation (3) indicates that the vectors from the same representation but from different sets of elements in the matrices of this representation are orthogonal.

Equation (4) indicates the fact that the square of the length of any such vector equals $\frac{h}{l_i}$.

The operators $\hat{C}_2, \hat{\sigma}_v, \hat{\sigma}'_v$ and $\hat{E}$ form a group:

Taking water molecule as the object of operations, let σ_v be the reflection in the molecular plane and σ'_v the reflection in the plane bisecting the *H—O—H* angle. Then constructing the multiplication table by investigating the effect of two successive operations.

The effect of $\hat{C}_2$ followed by another $\hat{C}_2$ is to bring the water molecule back to the original position.

$$\therefore \qquad \hat{C}_2 \hat{C}_2 = \hat{E}$$

The result of $\hat{\sigma}_v$ followed by $\hat{\sigma}'_v$ is the same as the rotation by 180°. Hence

$$\hat{\sigma}_v \hat{\sigma}'_v = \hat{C}_2$$

By proceeding along these lines, we get the following group multiplication table

First Operation	*Second Operation*			
	$\hat{E}$	$\hat{C}_2$	$\hat{\sigma}_v$	$\hat{\sigma}'_v$
$\hat{E}$	$\hat{E}$	$\hat{C}_2$	$\hat{\sigma}_v$	$\hat{\sigma}'_v$
$\hat{C}_2$	$\hat{C}_2$	$\hat{E}$	$\hat{\sigma}'_v$	$\hat{\sigma}_v$
$\hat{\sigma}_v$	$\hat{\sigma}_v$	$\hat{\sigma}'_v$	$\hat{E}$	$\hat{C}_2$
$\hat{\sigma}'_v$	$\hat{\sigma}'_v$	$\hat{\sigma}_v$	$\hat{C}_2$	$\hat{E}$

From this table we see that the four conditions for group multiplication are met. Hence the above operations form a group. In this particular case the operations also commute.

A hypothetical molecule A_3 with the geometry of an equilateral triangle. The basis set consisted of one function from each atom. Symmetry adapted functions (SAF), constructed:

The Molecule belongs to the C_3 group

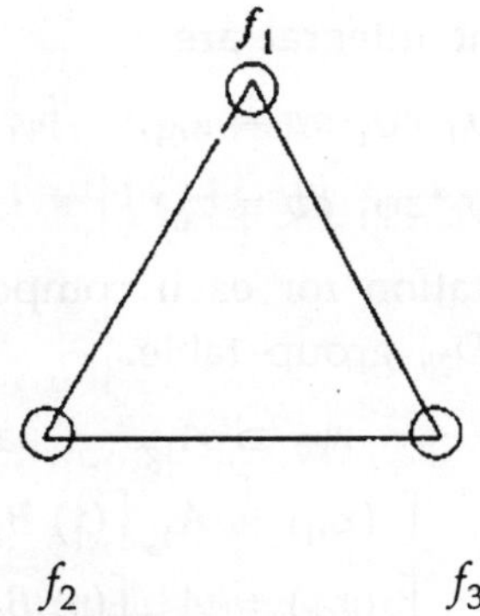

Let us denote the three basis functions as f_1, f_2 and f_3 and the SAF_s by $\psi(a)$, $\psi(e)$ and $\psi'(e)$.

From the projection operator theorem

$$\psi(a) = 1\hat{E}f_1 + 1\hat{C}_3 f_1 + 1\hat{C}_3^2 f_1 = f_1 + f_2 + f_3$$

The E representation is doubly degenerate.

The functions belonging to it are

$$\psi(e) = 2\hat{E}f_1 - 1\hat{C}_2 f_1 - 1\hat{C}_2^2 f_1 = 2f_1 - f_2 - f_3$$

By starting with the function f_2 we have

$$\psi'(e) = 2f_2 - f_3 - f_1$$

By starting with f_3, we obtain $(2f_3 - f_1 - f_2)$.

However, this function is not independent of the other two functions, as may be seen by adding them. Instead of the above set we may work with

$$\psi_1 = \psi(e) + \psi'(e) = 2f_3 - f_1 - f_2$$

and $\quad \psi_2 = \psi(e) - \psi'(e) = f_1 - f_2$

Since any linear combination of degenerate eigen functions is also an eigen function.

The $^1A_g \rightarrow {}^1B_{3u}$ transition in pyrazine is allowed:

For a transition to be allowed the transition moment integral must have a non-zero value. The components of the transition moment integral are

$$\int \psi_0 \, x\psi_1 \, dv \equiv x_{01}, \quad \int \psi_0 \, y\psi_1 \, dv = \psi_{01}$$

$$\int \psi_0 \, z\psi_1 \, dv \equiv z_{01}$$

The representation for each component is obtained by referring to the D_{2h} group table.

Since $\quad \psi_0 \supset A_{1g} \quad$ and $\quad \psi_1 \supset B_{3u}$

$$\Gamma(x_{01}) = A_{1g}\Gamma(x)\, B_{3u} = A_{1g}\, B_{3u}\, B_{3u} = A_{1g}$$

$$\Gamma(y_{01}) = A_{1g}\Gamma(y)\, B_{3u} = A_{1g}\, B_{2u}\, B_{3u} = B_{1g}$$

$$\Gamma(z_{01}) = A_{1g}\Gamma(z)\, B_{3u} = A_{1g}\, B_{1u}\, B_{3u} = B_{2g}$$

$\therefore \quad x_{01} \neq 0 \quad$ and $\quad y_{01} = z_{01} = 0$

We see from this that the transition is allowed in the direction perpendicular to the molecular plane.

Relation with Quantum Chemistry

The Principles of Symmetry and Group Theory used in the Area of Quantum Chemistry:

Principles of symmetry and group theory find applications in several areas of quantum chemistry like chemical bonding, molecular spectroscopy, ligand field theory, crystal field theory, etc. The procedure in all these cases involves—

(*i*) Generating a reducible representation Γr of the symmetry group to which the molecule belongs using a set of atomic orbitals as basis.

(*ii*) Resolving the Γr to irreducible representation Γi's using formula.

$$a_i = \frac{1}{h_{\text{all classes}}} \Sigma\chi(R)\,\chi_i\,(R)\,.\,n$$

$a_i \rightarrow$ number of irreducible representations (symmetry types) of one kind i in T_r, n is the total number of symmetry operations in the symmetry group. $\chi(R)$ is character of the operation R in the reducible representation, χ_i (R) is the character of the same operation R in the irreducible representation i, and n is number of the operations in one class.

(*iii*) Constructing SALC's (symmetry adapted linear combinations) corresponding to the T_i's of the group.

Energy Level Diagrams for:

(a) π orbitals of napthalene,

(b) Ferrocene.

(*a*) Napthalene belongs to the point group D_{2h}.

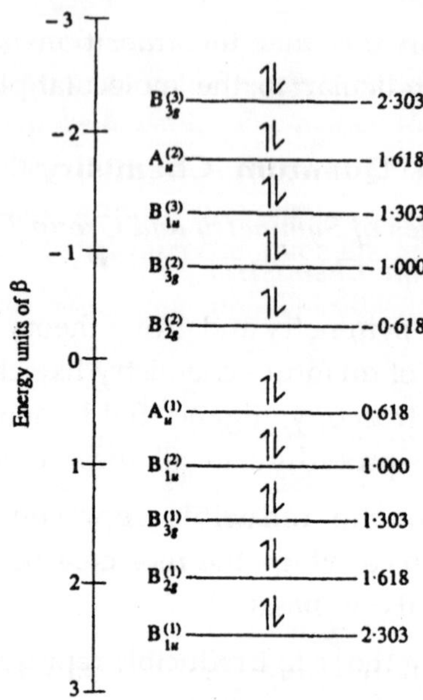

Energy level diagram for n orbitals of naphthalene.

(*b*) Ferrocene belongs to D_{5d} group.

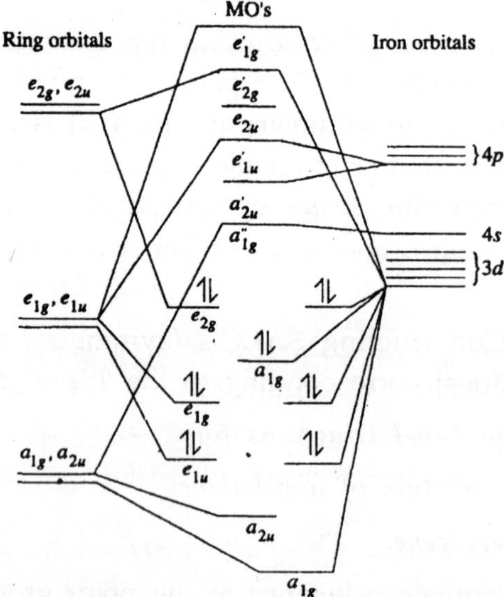

An energy level diagram for ferrocene.

Different Models

The important rules about irreducible representations:

(1) The sum of the squares of the dimensions of the irreducible representations of a group is equal to the order of the group, that is

$\Sigma l_1^2 = l_1^2 + l_2^2 + l_3^2 + \ldots\ldots = h$

(2) The sum of the squares of the characters in any irreducible representation equals *h,* that is

$$\sum_R \left[X_i(R)\right]^2 = h$$

(3) The vectors whose components are the characters of two different irreducible representations are orthogonal, that is

$$\sum_R X_i(R)\, X_j(R) = 0 \quad \text{when} \quad i \neq j$$

(4) In a given representation (reducible or irreducible) the characters of all matrices belonging to operations in the same class are identical.

(5) The number of the irreducible representations of a group is equal to the number of classes in the group.

The energy level diagram of the molecular orbitals in the octahedral symmetry and forming only a bonds, drawn:

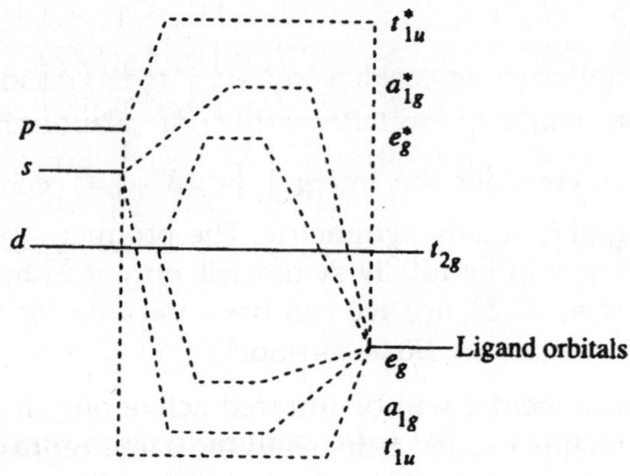

Since only particular kinds of orbitals lead to octahedral symmetry, *i.e.*, $d_{x^2-y^2}$, d_{z^2}, s, p_x, p_y, p_z so the d_{xy}, d_{yz} and d_{xz} will be unaffected by the bonding as these orbitals are not involved in the bond formation. The Δ, splitting between T_{2g} & e_g^* depends upon the strength of the ligands.

The molecular orbital diagram for $[FeF_6]^{3-}$ complex, drawn:

Since F^- ion is a weak ligand, the electrostatic field splitting Δ will be small and will lead to high spin complex. The molecular orbital energy level diagram will be

t^*_{1u}	—		—		—
a^*_{1g}			—		
e^*_g		↑		↑	
t_{2g}	↑		↑		↑
e_g		xx		xx	
a_{1g}			xx		
t_{1u}	xx		xx		xx

Note: xx → Indicates the electrons of six ligands.

The selection rule for IR spectra:

For a fundamental transition to occur by absorption of Infrared radiation, it is necessary that one of integrals of equation (1) given below be non-zero.

$$\left.\begin{aligned} I_x &\propto \int \psi_i \, X \, \psi_j \, d\xi \\ I_y &\propto \int \psi_i \, Y \, \psi_j \, d\xi \\ I_z &\propto \int \psi_i \, Z \, \psi_j \, d\xi \end{aligned}\right\} \qquad \ldots(1)$$

Let us consider the integral $\int \psi_i \, X \, \psi_j \, d\xi$ consider ψ_i in this integral is totally symmetric. The product representation of X and ψ_i will be totally symmetric only if ψ_j has the same symmetry as X. Hence we can have the rule for the activity of fundamentals in *IR* absorption.

A fundamental will be infrared active only if the normal mode which is excited belongs to the same representation as any one or several of the Cartesian coordinates.

The HOMO in CO is:

(a) π bonding ***(b) σ bonding***

(c) π antibonding ***(d) σ antibonding***

'd' because *MO* configuration of *CO* is $(1\sigma)^2$ $(2\sigma)^2$ $(1\pi)^4$ $(3\sigma)^2$. The HOMO is 3σ orbital which is antibonding.

The rules of irreducible representation for C_{2v} point group:

The C_{2v} consist of 4 element and each is in a separate class. Hence according to rule there will be four irreducible representations for the group and according to rule the sum of the squares of dimensions of these representations equals to *g*. Thus

$$l_1^2 + l_2^2 + l_3^2 + l_4^2 = 4$$

Thus $$l_1 = l_2 \pm l_3 = l_4 = 1$$

Thus C_{2v} has four one-dimensional irreducible representation.

On working out the characters of these four irreducible representations on the basis of vector properties and rules of irreducible representations one can write,

C_{2v}	E	C_2	σ_v	σ'_v
$\Gamma 1$	1	1	1	1

This is a suitable vector in 4-space which has a component of 1 corresponding to *E*.

From this $$\Sigma[X, (R)]^2 = 1^2 + 1^2 + 1^2 + 1^2 = 4$$

thus satisfying the rule.

Now all other representations are to be such that

$$\sum_R [X_1(R)]^2 = 4$$

which will be true for $$X_1(R) = \pm 1$$

According to rule in order for each of the other representation to be orthogonal to $\Gamma 1$ there must be two + 1 and two – 1.

Thus $(1)(-1) + (1) + (-1) + (1)\,(1) + (1)\,(1) = 0$

The total can be written as

C_{2v}	E	C_2	σ_v	σ'_v
$\Gamma 1$	1	1	1	1
$\Gamma 2$	1	–1	–1	1
$\Gamma 3$	1	–1	1	–1
$\Gamma 4$	1	1	–1	–1

Trigonal planar molecule such as BF_3 cannot have degenerate orbitals:

BF_3 belongs to D_{3h} point group. The maximum number in the column headed by the identity E is the maximum orbital degeneracy possible in a molecule of that symmetry group. The character table of D_{3h} shows that maximum degeneracy is 2, as no character exceeds 2 in the column headed E. This means, the orbitals cannot be triply degenerate.

The multiplication table of the Pauli Spin matrices and the 2 × 2 unit matrix.

$$\sigma_x = \begin{pmatrix} 0 & 1 \\ 1 & 0 \end{pmatrix} \qquad \sigma_y = \begin{pmatrix} 0 & -i \\ -i & 0 \end{pmatrix}$$

$$\sigma_z = \begin{pmatrix} 1 & 0 \\ 0 & -1 \end{pmatrix} \qquad E = \begin{pmatrix} 1 & 0 \\ 0 & 1 \end{pmatrix}$$

The multiplication table is—

	E	σ_x	σ_y	σ_z
E	E	σ_x	σ_y	σ_z
σ_x	σ_x	E	$i\sigma_z$	$-i\sigma_y$
σ_y	σ_y	$-i\sigma_y$	E	$i\sigma_x$
σ_z	σ_z	$i\sigma_y$	$-\sigma_x$	E

The matrices do not form a group since the product $1\sigma_z$, $i\sigma_y$, $i\sigma_x$ and their negatives are not among the four given matrices.

Taken 1s orbitals as a basis of the two hydrogen and the four valence orbitals of the oxygen atom in H_2O molecule to set up 6 × 6 matrices and then confirm by explicit matrix multiplication the group multiplication (i) $C_2T_V = T'_v$ and (ii) $\sigma_v\ \sigma'_v = C_2$.

H_2O belongs to C_{2v} point group.

Places orbitals h_1 and h_2 on the H atoms and s, p_x, p_y and p_z on the O atom. The z-axis is the C_2 axis, X lies perpendicular to σ'_v, y lies perpendicular to σ_v. Then draw up the following table of the effect of the operations on the basis—

	E	C_2	σ_v	σ'_v
h_1	h_1	h_2	h_2	h_1
h_2	h_2	h_1	h_1	h_2
s	s	s	s	s
p_x	p_x	$-p_x$	p_x	$-p_x$
p_y	p_y	$-p_y$	$-p_y$	p_y
p_z	p_z	p_z	p_z	p_z

Express the column headed by each operation R in the form (new) = $D(R)$ original, where $D(R)$ is the 6 × 6 representative of the operation R. We use the rules of matrix multiplication.

E: $(h_1, h_2, s, p_x, p_y, p_z) \leftarrow (h_1, h_2, s, p_x, p_y, p_z)$ is reproduced by the 6 × 6 unit matrix.

C_2: $(h_2, h_1, s, -p_x, -p_y, p_z) \leftarrow (h_1, h_2, s, p_x, p_y, p_z)$ is reproduced by

$$D(C_2) = \begin{bmatrix} 0 & 1 & 0 & 0 & 0 & 0 \\ 1 & 0 & 0 & 0 & 0 & 0 \\ 0 & 0 & 1 & 0 & 0 & 0 \\ 0 & 0 & 0 & -1 & 0 & 0 \\ 0 & 0 & 0 & 0 & -1 & 0 \\ 0 & 0 & 0 & 0 & 0 & -1 \end{bmatrix}$$

$\sigma_v = (h_2, h_1, s, p_x, -p_y, p_z) \leftarrow (h_1, h_2, s, p_x, p_y, p_z)$ is reproduced by

$$D(\sigma_v) = \begin{bmatrix} 0 & 1 & 0 & 0 & 0 & 0 \\ 1 & 0 & 0 & 0 & 0 & 0 \\ 0 & 0 & 1 & 0 & 0 & 0 \\ 0 & 0 & 0 & 1 & 0 & 0 \\ 0 & 0 & 0 & 0 & -1 & 0 \\ 0 & 0 & 0 & 0 & 0 & 1 \end{bmatrix}$$

$\sigma'_v = (h_1, h_2, s, -p_x, p_y, p_z) \leftarrow (h_1, h_2, s, p_x, p_y, p_z)$ is reproduced by

$$D(\sigma'_v) = \begin{bmatrix} 1 & 0 & 0 & 0 & 0 & 0 \\ 0 & 1 & 0 & 0 & 0 & 0 \\ 0 & 0 & 1 & 0 & 0 & 0 \\ 0 & 0 & 0 & -1 & 0 & 0 \\ 0 & 0 & 0 & 0 & 1 & 0 \\ 0 & 0 & 0 & 0 & 0 & 1 \end{bmatrix}$$

(*i*) To confirm the correct representation of $C_2\sigma_v = \sigma'_v$, we write $D(C_2)\,D(\sigma_v)$

$$= \begin{bmatrix} 0 & 1 & 0 & 0 & 0 & 0 \\ 1 & 0 & 0 & 0 & 0 & 0 \\ 0 & 0 & 1 & 0 & 0 & 0 \\ 0 & 0 & 0 & -1 & 0 & 0 \\ 0 & 0 & 0 & 0 & 1 & 0 \\ 0 & 0 & 0 & 0 & 0 & 1 \end{bmatrix} \begin{bmatrix} 0 & 1 & 0 & 0 & 0 & 0 \\ 1 & 0 & 0 & 0 & 0 & 0 \\ 0 & 0 & 1 & 0 & 0 & 0 \\ 0 & 0 & 0 & 1 & 0 & 0 \\ 0 & 0 & 0 & 0 & -1 & 0 \\ 0 & 0 & 0 & 0 & 0 & 1 \end{bmatrix}$$

$$= \begin{bmatrix} 1 & 0 & 0 & 0 & 0 & 0 \\ 0 & 1 & 0 & 0 & 0 & 0 \\ 0 & 0 & 1 & 0 & 0 & 0 \\ 0 & 0 & 0 & -1 & 0 & 0 \\ 0 & 0 & 0 & 0 & 1 & 0 \\ 0 & 0 & 0 & 0 & 0 & 1 \end{bmatrix} = D(\sigma'_v)$$

(*ii*) Similarly, to confirm the correct representation of σ_v $\sigma'_v = C_2$, we write

$$\begin{bmatrix} 0 & 1 & 0 & 0 & 0 & 0 \\ 1 & 0 & 0 & 0 & 0 & 0 \\ 0 & 0 & 1 & 0 & 0 & 0 \\ 0 & 0 & 0 & 1 & 0 & 0 \\ 0 & 0 & 0 & 0 & -1 & 0 \\ 0 & 0 & 0 & 0 & 0 & 1 \end{bmatrix} \begin{bmatrix} 1 & 0 & 0 & 0 & 0 & 0 \\ 0 & 1 & 0 & 0 & 0 & 0 \\ 0 & 0 & 1 & 0 & 0 & 0 \\ 0 & 0 & 0 & -1 & 0 & 0 \\ 0 & 0 & 0 & 0 & 1 & 0 \\ 0 & 0 & 0 & 0 & 0 & 1 \end{bmatrix}$$

$$= \begin{bmatrix} 0 & 1 & 0 & 0 & 0 & 0 \\ 1 & 0 & 0 & 0 & 0 & 0 \\ 0 & 0 & 1 & 0 & 0 & 0 \\ 0 & 0 & 0 & -1 & 0 & 0 \\ 0 & 0 & 0 & 0 & -1 & 0 \\ 0 & 0 & 0 & 0 & 0 & -1 \end{bmatrix} = D(C_2)$$

The irreducible components of representations generated by a set of a type atomic orbitals in XY_3 molecules of C_{3v} and D_{3h} symmetry.

C_{3v}:

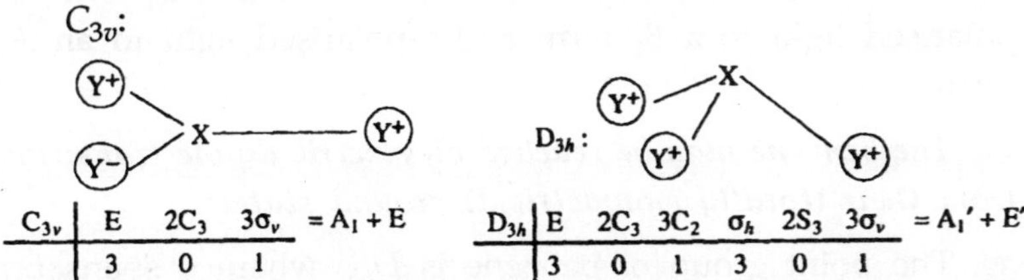

C_{3v}	E	$2C_3$	$3\sigma_v$	$= A_1 + E$
	3	0	1	

D_{3h}	E	$2C_3$	$3C_2$	σ_h	$2S_3$	$3\sigma_v$	$= A_1' + E'$
	3	0	1	3	0	1	

In case methane molecule is distorted to (a) C_{3v}, point group due to bond lengthening and (b) to C_{2v} point group due to scissor action of molecular vibration. More d-orbitals would become available for bonding.

(*a*) In C_{3v} symmetry the *H*1*s* orbitals span the same irreducible representations as in NH_3, which is $A_1 + A_2 + E$. There is an additional A_1 orbital because a fourth *H* atom lies on the C_3 axis. In C_{3v}, the *d* orbitals span $A_1 + E + E$. Therefore, all five *d* orbitals may contribute to the bonding.

(*b*) In C_{2v} symmetry the $H1s$ orbitals span the same irreducible representations as in H_2O, but one 'H_2O' fragment is rotated by 90° with respect to the other. Therefore, whereas in H_2O the $H1s$ orbitals span $A_1 + B_2$ $[H_1 + H_2, H_1 - H_2]$ in the distorted CH_4 molecule they span $A_1 + B_2 + A_1 + B_1$ $[H_1 + H_2, H_1 - H_2, H_3 + H_4, H_3 - H_4]$. In C_{2v} the *d*-orbitals span $2A_1 + B_1 + B_2 + A_2$, therefore, all except A_2 (d_{xy}) may participate in bonding.

The ground state of NO_2 (C_{2v}) is A_1. The excited states may be excited by electric dipole transition and whose polarisation of light is it necessary to use:

Ans. The electric dipole moment operator transforms as x (B_1), y (B_2) and z (A_1) [C_{2v} character table]. Transitions are allowed if $\int \psi_f \, \mu \, \psi_i \, d\tau$ is non-zero and hence are forbidden unless $\Gamma f \times \Gamma(\mu) \times \Gamma i$ contains A_1.

Since $\Gamma i = A_1$, this requires $\Gamma f \times \Gamma(\mu) = A_1$

Since $B_1 \times B_1 = A_1$, and $B_2 \times B_2 = A_1$ and $A_1 \times A_1 = A_1$

x-polarised light may cause a transition to a B_1 term, y-polarised light to a B_2 term and z-polarised light to an A_1 term.

The benzene may be reached by electric dipole transition from their (totally symmetrical) ground states:

The point group of benzene is D_{6h}, where μ spans E_{1u} (x, y) and A_{2u} (z) and the group term is A_{1g}. Then using $A_{2u} \times A_{1g} = A_{2u}$, $E_{1u} \times A_{1g} = E_{1u}$, $A_{2u} \times A_{2u} = A_{1g}$ and $E_{1u} \times E_{1u} = A_{1g} + A_{2g} + E_{2g}$ we conclude that the upper term is either E_{1u} or A_{2u}.

The symmetry elements listed and named the point groups to which following molecule belongs:

(i) Staggered CH_3CH_3, (ii) Chair and boat cyclohexane, (iii) B_2H_6, (iv) $[Co(en)_3]^{3+}$, (v) S_8 which of them molecules can be polar and chiral.

(*i*) Staggered CH_3CH_3: $E, C_3, C_2, 3\sigma_d; D_3d$

(*ii*) Chair $C_6H_{12:}$ $E, C_3, C_2, 3\sigma_{d;}$ D_{3d}

Boat C_6H_{12}: $E, C_2, \sigma_v, \sigma'_v, C_{2v}$

(*iii*) $B_2H_{6:}$ $E, C_2, 2\ C'_2; \sigma_n; D_{2h}$

(*iv*) $[Co(en)_3]^{3+}$: $E, 2C_3, 3C_2; D_3$

(*v*) Crown $S_{8:}$ $E, C_4, C_2, 4C'_2, 4\sigma_d, 2S_8; D_{4d}$

Only boat C_6H_{12} may be polar, since all the others are D point groups. Only $[Co(en)_3]^{3+}$ belongs to a group without an improper rotation axis ($S_1 = \sigma$) and hence is chiral.

A weak infrared band is seen at 2903 cm^{-1} in the IR spectrum of BF$_3$. The feature assigned and deduced its symmetry properties:

The band can be assigned to $2v_3$, the first overtone of the E' B—F stretch (2 × 1453). To determine the symmetry species of this overtone use the equation written below

$$\chi_2(R) = \frac{1}{2}\left[\chi(R)\,\chi(R) + \chi\left(R^2\right)\right]$$

For D_{3h}, we have

R	E	C_3^1	C_3^2	C_2	C_2	C_2	σ_1	S_3^1	S_3^5	σ_v	σ_v	σ_v
R^2	E	C_3^2	C_3^1	E	E	E	5	C_3^2	C_3^1	E	E	E
$\chi(R)$	2	−1	−1	0	0	0	2	1	−1	0	0	0
$\chi(R^2)$	2	−1	−1	2	2	2	2	−1	−1	2	2	2
$\chi_2(R)$	3	0	0	1	1	1	3	0	0	1	1	1

The resulting representation is reducible to $A_1 + E'$. Hence, the first overtone of the E' B—F stretch of BF_3 has symmetry species A'_1 and E'.

The transition $A_1 \rightarrow A_2$ is forbidden for electric dipole transmission is C_{3v} molecule:

Considering all three components of the electric dipole moment operator, μ.

Component of μ:	X			Y			Z		
A_1	1	1	1	1	1	1	1	1	1
$\Gamma(\mu)$	2	–1	0	2	–1	0	1	1	1
A_2	1	1	–1	1	1	–1	1	1	–1
$A_1\Gamma(\mu)A_2$	2	–1	0	2	–1	0	1	1	–1
		E			E			A_2	

Since A_1 is not present in any product, the transition dipole moment must be zero.

Raman Spectra

The selection rule for Raman spectra:

For Raman scattering it is necessary that atleast one of the integrals of the type $\int \psi_v^0 \, P\psi_v \, / \, d\xi$ be non-zero. In these types of integrals P is one of the quadratic function of the cartesian coordinates namely, x^2, y^2, z^2, xy, yz, zx all of which are listed opposite to the representations that they generate in the character tables. These P's are components of the polarizability tensor. The above integral will become non-zero only if there is a change in polarizability of the molecule during transition. A fundamental transition will be Raman active only if the normal mode involved belongs to the same representation as one or more of the components of the polarizability tensor of the molecule. For example for NH_3 molecule the character table for C_{3v} group is used to obtain the following irreducible representations for the quadratic and binary cartesian coordinates.

Quadratic and Binary Cartesian Coordinates	*Representation*
Z^2	A_1
$(x^2 - y^2, xy)$ (xz, yz)	E

The representations obtained for the quadratic and binary coordinates [Z^2; $(x^2 - y^2, xy)$; (xz, yz)] correspond to the symmetry species of the vibrational modes for NH_3 molecule. Therefore, the vibrational modes of NH_3 molecule are Raman active.

Experimental and Observational Studies

A common goal for a statistical research project is to investigate causality, and in particular to draw a conclusion on the effect of changes in the values of predictors or independent variables or dependent variables on response. There are two major types of causal statistical studies: experimental studies and observational studies. In both types of studies, the effect of differences of an independent variable (or variables) on the behaviour of the dependent variable are observed. The difference between the two types lies in how the study is actually conducted. Each can be very effective.

An experimental study involves taking measurements of the system under study, manipulating the system, and then taking additional measurements using the same procedure to determine if the manipulation has modified the values of the measurements. In contrast, an observational study does not involve experimental manipulation. Instead, data are gathered and correlations between predictors and response are investigated.

An example of an experimental study is the famous Hawthorne studies, which attempted to test the changes to the working environment at the Hawthorne plant of the Western Electric Company. The researchers were interested in determining whether increased illumination would increase the productivity of the assembly line workers.

The researchers first measured the productivity in the plant, then modified the illumination in an area of the plant and checked if the changes in illumination affected the productivity. It turned out that the productivity indeed improved (under the experimental conditions). However, the study is heavily criticised today for errors in experimental procedures, specifically for the lack of a control group and blindness.

An example of an observational study is a study which explores the correlation between smoking and lung cancer. This type of study typically uses a survey to collect observations about the area of interest and then performs statistical analysis. In this case, the researchers would collect observations of both smokers and non-smokers, perhaps through a case-control study, and then look for the number of cases of lung cancer in each group.

The basic steps of an experiment are:

- Planning the research, including determining information sources, research subject selection, and ethical considerations for the proposed research and method.
- Design of experiments, concentrating on the system model and the interaction of independent and dependent variables.
- Summarising a collection of observations to feature their commonality by suppressing details. (Descriptive statistics).

- Reaching consensus about what the observations tell about the world being observed. (Statistical inference).
- Documenting / presenting the results of the study.

Levels of Measurement

There are four types of measurements or levels of measurement or measurement scales used in statistics: nominal, ordinal, interval and ratio. They have different degrees of usefulness in statistical research. Ratio measurements have both a zero value defined and the distances between different measurements defined; they provide the greatest flexibility in statistical methods that can be used for analysing the data. Interval measurements have meaningful distances between measurements defined, but have no meaningful zero value defined (as in the case with IQ measurements or with temperature measurements in Fahrenheit). Ordinal measurements have imprecise differences between consecutive values, but have a meaningful order to those values. Nominal measurements have no meaningful rank order among values.

Since variables conforming only to nominal or ordinal measurements cannot be reasonably measured numerically, sometimes they are called together as *categorical variables,* whereas ratio and interval measurements are grouped together as quantitative or continuous variables due to their numerical nature.

Statistical Techniques

Some well-known statistical tests and procedures are:

- Student's t-Test.
- Chi-square Test.
- Analysis of Variance (ANOVA).
- Mann-Whitney U.

- Regression Analysis.
- Factor Analysis.
- Correlation.
- Pearson Product-moment Correlation Coefficient.
- Spearman's Rank Correlation Coefficient.
- Time Series Analysis.

Specialised Disciplines

Some fields of enquiry use applied statistics so extensively that they have specialised terminology. These disciplines include:

- Actuarial Science.
- Applied Information Economics.
- Biostatistics.
- Bootstrapand Jackknife Resampling.
- Business Statistics.
- Data Analysis.
- Data Mining (applying statistics and pattern recognition to discover knowledge from data).
- Demography.
- Economic Statistics (Econometrics).
- Energy Statistics.
- Engineering Statistics.
- Environmental Statistics.
- Epidemiology.
- Geography and Geographic Information Systems, Specifically in Spatial Analysis.
- Image Processing.

- Psychological Statistics.
- Quality.
- Social Statistics.
- Statistical Literacy.
- Statistical Modelling.
- Statistical Surveys.
- Chemometrics (for analysis of data from chemistry).
- Structured Data Analysis (statistics).
- Survival Analysis.
- Reliability Engineering.
- Statistics in Various Sports, Particularly Baseball and Cricket

Statistics form a key basis tool in business and manufacturing as well. It is used to understand measurement systems variability, control processes (as in statistical process control or SPC), for summarising data, and to make data-driven decisions. In these roles, it is a key tool, and perhaps the only reliable tool.

Statistical Computing

The rapid and sustained increases in computing power starting from the second half of the 20th century have had a substantial impact on the practice of statistical science. Early statistical models were almost always from the class of linear models, but powerful computers, coupled with suitable numerical algorithms, caused an increased interest in non-linear models (especially neural networks) as well as the creation of new types, such as generalised linear models and multilevel models.

Increased computing power has also led to the growing popularity of computationally-intensive methods based on resampling, such as permutation tests and the bootstrap, while techniques such as Gibbs sampling have made use of Bayesian models more feasible. The computer revolution has implications for the future of statistics with new emphasis on "experimental" and "empirical" statistics. A large number of both general and special purpose statistical software are now available.

Misuse of Statistics

A misuse of statistics occurs when a statistical argument asserts a falsehood. In some cases, the misuse may be accidental. In others, it is purposeful and for the gain of the perpetrator. When the statistical reason involved is false or misapplied, this constitutes a statistical fallacy.

The false statistics trap can be quite damaging to the quest for knowledge. For example, in medical science, correcting a falsehood may take decades and cost lives.

Misuses can be easy to fall into. Professional scientists, even mathematicians and professional statisticians, can be fooled by even some simple methods, even if they are careful to check everything. Scientists have been known to fool themselves with statistics due to lack of knowledge of probability theory and lack of standardisation of their tests.

Discarding Unfavourable Data

In product quality control terms all a company has to do to promote a neutral (useless) product is to find or conduct, for example, 40 studies with a confidence level of 95 per cent. If the product is really useless, this would on average produce one study showing the product was beneficial, one study showing it was harmful and thirty-eight inconclusive studies (38 is 95 per cent of 40). This tactic becomes more effective the more studies there are available. Organisations that do not publish every study they carry out, such as tobacco companies denying a link between smoking and cancer, or miracle pill vendors, are likely to use this tactic.

Another common technique is to perform a study that tests a large number of dependent (response) variables at the same time. For example, a study testing the effect of a medical treatment might use as dependent variables the probability of survival, the average number of days spent in the hospital, the patient's self-reported level of pain, etc. This also increases the likelihood that at least one of the variables will by chance show a correlation with the independent (explanatory) variable.

Loaded Questions

The answers to surveys can often be manipulated by wording the question in such a way as to induce a prevalence towards a certain answer from the respondent. For example, in polling support for a war, the questions:

- Do you support the attempt by the war-making country to bring freedom and democracy to other places in the world?
- Do you support the unprovoked military action by the war-making country?

These questions will likely result in data skewed in different directions, although they are both polling about the support for the war.

Another way to do this is to precede the question by information that supports the "desired" answer. For example, more people will likely answer "yes" to the question, "Given the increasing burden of taxes on middle-class families, do you support cuts in income tax?" than to the question, "Considering the rising federal budget deficit and the desperate need for more revenue, do you support cuts in income tax?"

Overgeneralisation

If you have a statistic saying 100 per cent of apples are red in summer, and then publish, "All apples are red", you will be overgeneralising because you only looked at apples in summertime and are using that data to make inferences about apples in all seasons. Often, the overgeneralisation is made not by the original researcher, but by others interpreting the data. Continuing with the previous example, if on a TV show you say, "All apples are red in summer" many people will not remember you said, "in summer" if asked weeks later.

With a subject of which the general public has no personal knowledge, you can fool a lot of people. For example you can say on TV, "Most autistics are hopelessly incurable if raised without parents or normal education" and many people will only remember the first part of the claim, "Most autistics are hopelessly incurable". This problem is especially prevalent on TV, where talk show hosts interview one individual as representative of a whole class of people.

Misreporting or Misunderstanding of Estimated Error

If a research team wants to know how 300 million people feel about a certain topic, it would be impractical to ask all of them. However, if the team picks a random sample of about 1,000 people, they can be fairly certain that the results given by this group are representative of what the larger group would have said if they had all been asked.

This confidence can actually be quantified by the central limit theorem and other mathematical results. Confidence is expressed as a probability of the true result (for the larger group) being within a certain range of the estimate (the figure for the smaller group). This is the "plus or minus" figure often quoted for statistical surveys. The probability part of the confidence level is usually not mentioned; if so, it is assumed to be a standard number like 95 per cent.

The two numbers are related. If a survey has an estimated error of ±5 per cent at 95 per cent confidence, it might have an estimated error of ±6.6 per cent at 99 per cent confidence. The larger the sample, the smaller the estimated error at a given confidence level.

Most people assume, because the confidence figure is omitted, that there is a 100 per cent certainty that the true result is within the estimated error. This is not mathematically correct.

Many people may not realise that the randomness of the sample is very important. In practice, many opinion polls are conducted by phone, which distorts the sample in several ways, including exclusion of people who do not have phones, favouring the inclusion of people who have more than one phone, favouring the inclusion of people who are willing to participate in a phone survey over those who refuse, etc. Non-random sampling makes the estimated error unreliable.

On the other hand, many people consider that statistics are inherently unreliable because not everybody is called, or because they themselves are never polled. Many people think that it is impossible to get data on the opinion of dozens of millions of people by just polling a few thousands. This is also inaccurate. A poll with perfect unbiased sampling and truthful answers has a mathematically determined margin of error, which only depends on the number of people polled.

Another problem that crops up is the "resampling" problem. For example, a survey of 1,000 people may contain 100 people from a certain ethnic or economic group. The people taking the survey would then "resample" their results focusing on that group. They then make claims that a percentage of that group believes X, or whatever the survey is about.

Unfortunately, resampling reduces the statistical reliability of the data. The larger the total sample is, then the more likely the sample represents the population. So, consider the case where a sampling is done and has a margin of error of only 3 per cent. Those reporting the statistics from the sample can say the answers to the questions represent population results plus or minus 3 per cent.

However, that error rate does not hold to subsets of the samples. If you take a subgroup that uses only 10 per cent of the samples (say 100 samples from a 1,000-sample finding) the error grows significantly higher than the original 3 per cent. To accurately find statistics for a subgroup, that subgroup would have to be sampled by itself (e.g. the sub group would need the same number of samples as the original group).

Resampling errors seem to occur all the time in news coverage of survey data.

The problems mentioned above apply to all statistical experiments, not just population surveys.

There are also many other measurement problems in population surveys.

False Causality

When a statistical test shows a correlation between A and B, there are usually four possibilities:

1. A causes B.
2. B causes A.
3. A and B are both caused by a third factor, C.
4. The observed correlation was due purely to chance.

The fourth possibility can be quantified by statistical tests that can calculate the probability that the correlation observed would be as large as it is just by chance if, in fact, there is no relationship between the variables. However, even if that possibility has a small probability, there are still the three others.

If the number of people buying ice cream at the beach is statistically related to the number of people who drown at the beach, then nobody would claim ice cream causes drowning because it's obvious that it isn't so. (In this case, both drowning and ice cream buying are clearly related by a third factor: the number of people at the beach).

This fallacy can be used, for example, to prove that exposure to a chemical causes cancer. Replace "number of people buying ice cream" with "number of people exposed to chemical X", and "number of people who drown" with "number of people who get cancer", and many people will believe you. In such a situation, there may be a statistical correlation even if there is no real effect.

For example, if there is a perception that the chemical is "dangerous" (even if it really isn't) property values in the area will decrease, which will entice more low-income families to move to that area. If low-income families are more likely to get cancer than high-income families (this can happen for many reasons, such as a poorer diet or less access to medical care) then rates of cancer will go up, even though the chemical itself is not dangerous. It is believed that this is exactly what happened with some of the early studies showing a link between EMF (electromagnetic fields) from power lines and cancer.

In well-designed studies, the effect of false causality can be eliminated by assigning some people into a "treatment group" and some people into a "control group" at random, and giving the treatment group the treatment and not giving the control group the treatment. In the above example, a researcher might expose one group of people to chemical X and leave a second group unexposed. If the first group had higher cancer rates, the researcher knows that there is no third factor that affected whether a person was exposed because he controlled who was exposed or not, and he assigned people to the exposed and non-exposed groups at random.

However, in many applications, actually doing an "experiment" in this way is either prohibitively expensive, infeasible, unethical, illegal, or downright impossible. (For example, it is highly unlikely that an IRB would accept an experiment that involved intentionally exposing people to a dangerous substance in order to test its toxicity).

Data Dredging

Data dredging is an abuse of data mining. In data dredging, large compilations of data are examined in order to find a correlation, without any predefined choice of a hypothesis to be tested. Since the required confidence interval to establish a relationship between two parameters is usually chosen to be 95 per cent (meaning that there is a 95 per cent chance that the relationship observed is not due to random chance), there is a thus a 5 per cent chance of finding a correlation between any two sets of completely random variables. Given that data dredging efforts typically examine large datasets with many variables, and hence even larger numbers of pairs of variables, spurious but apparently statistically significant results are almost certain to be found by any such study.

Note that data dredging is a valid way of finding a possible hypothesis but that hypothesis must then be tested with data not used in the original dredging. The misuse comes in when that hypothesis is stated as fact without further validation.

Data Manipulation

Data manipulation is the presentation of scientific data in a misleading way to support a hypothesis which is actually without merit. Informally called "fudging the data", this practice includes selective reporting and even simply making up false data.

Examples of selective reporting abound. The easiest and most common examples involve choosing a group of results which follow a pattern consistent with the preferred hypothesis — while ignoring other results or "data runs" which contradict the hypothesis.

Psychic researchers have long disputed over studies showing people with ESP ability. Critics accuse ESP proponents of only publishing experiments with positive results and shelving those which show negative results. A "positive result" is a test run (or data run) in which the subject guesses a hidden card, etc., at a much higher frequency than random chance.

The deception involved in both cases is that the hypothesis is not confirmed by the totality of the experiments — only by a tiny, selected group of "successful" tests.

Scientists in general question the validity of study results which cannot be reproduced by other investigators. However, some scientists refuse to publish their data and methods.

Linguistically Asserting Unit Measure when it is Empirically Violated

Unit measure is an axiom of probability theory which states that, when an event is certain to occur, its probability is 1. This axiom is consistent with the empirical world, if the relation from a set of events that are certain to occur to a set of physical objects is one-to-one, but not otherwise. In the latter case, unit measure is scientifically invalidated.

Christensen and Reichert (1976), Oldberg and Christensen (1995) and Oldberg (2005) report observations of systems in which the relation is not one-to-one. A result of the lack of one-to-one-ness is that the following elements of statistical terminology are not defined for the associated systems:

1. Population,
2. Sampling unit,

3. Sample,
4. Probability and
5. Any term that assumes probability theory. A misuse of statistics arises when any of these terms are used in reference to a system that lacks one-to-one-ness, for unit measure is linguistically asserted and empirically violated. Oldberg and Christensen (1995) and Oldberg (2005) report observations of this type of misuse.

Significance of the Phenomenon

In Physical Chemistry, Thermo-electrical phenomenon are those in which heat energy is converted into electric energy or vice-versa. If the interchange can proceed in either direction (*i.e.* heat energy $\rightleftharpoons$ electric energy), then it (thermoelectric phenomenon) is said to be reversible, otherwise it is irreversible.

When an electric current is passed through a resistor of resistance R, then electrical energy is converted into heat energy. This effect is called *Joule effect.* The heat so produced is given by $\frac{i^2Rt}{J}$ kilo cal. and is called *Joule heat.* Joule effect proceeds in one direction only; reversal of the current has no effect upon it and it cannot be used to transform heat energy into electrical energy.

The reversible thermo-electric phenomenon are three in number. These occur at junctions of dissimilar conductors and/or throughout conductors that possess a temperature gradient. In the order of their discovery, they are called the *Seebeck effect*, the *Peltier effect* and the *Thomson effect*. The term thermo electric effect or *thermo-electricity*, customarily refers only to these reversible effects.

Power of Thermo-electricals

If the cold junction of a thermo-couple is kept at constant temperature and the temperature of the hot junction is changed, the thermo-electric e.m.f. of the couple also changes. The rate of change of thermo-e.m.f with temperature of the hot junction (*i.e.* dE/dT) is called the thermo-electric power of the couple at the particulr temperature. The thermo-electric power at a particular temperature is given by the slope of the tangent to the *E* – *T* curve at that temperature.

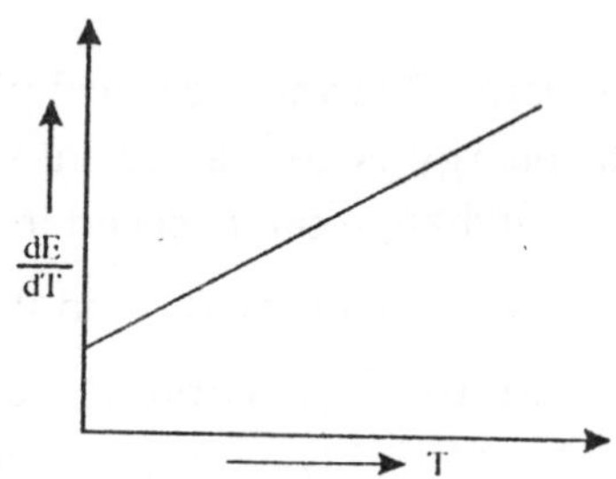

In order to determine the variation of thermo-electric power *dE/dT* with the temperature of the hot junction *T*, first of all a graph is plotted between the thermo-e.m.f. E and the temperature of the hot junction *T*. The graph is a parabola as already shown in fig. Now we choose different points on the parabola and draw tangents to the parabola on these points. The slopes of these tangents gives the values of the thermo-electric power *dE/dT* at temperature. *T* corresponding to these points. From these values a graph is plotted between thermoelectric power *dE/dT* and the temperature of the hot junction *T*.

Since the graph between thermo-electric e.m.f. E, and the temperature of the hot junction T is a parabola, therefore, when the temperature of the cold junction is kept at 0°C, the relation between thermo-electric e.m.f. E and temperature of the hot junction T is expressed as

$$E = aT + bT^2 \quad \text{(equation of parabola)}$$

where a and b are constants for the given thermo-couple.

Therefore, *the thermo-electric power* dE/dT *is* given by

$$\frac{dE}{dT} = a + 2bT.$$

which is the equation of a straight line ($y = mx + c$). Hence the graph between the thermo-electric power dE/dT and the temperature of the hot junction T is a straight line not passing through the origin as shown in fig.

Application of Thermodynamics on Thermo-couple: Since Peltier and Thomson effects are reversible, therefore, second law of thermocynamics can be applied to a thermo-couple circuit.

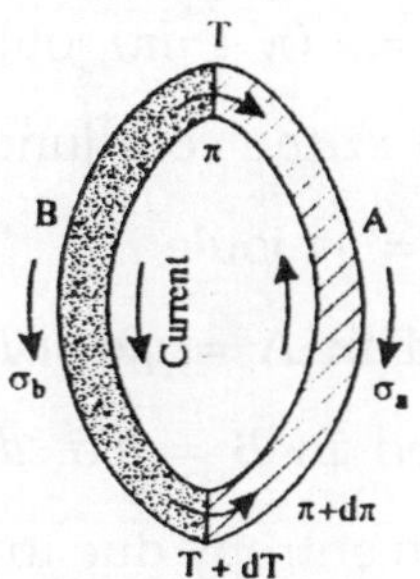

Let us take a thermo-couple of two metals A and B with its cold and hot junctions at absolute tempratures T and $(T + dT)$ respectively. Let π and $\pi + d\pi$ be the Peltier e.m.f.'s from B to A through the junctions at temperatures T and $T + dT$ respectively. Suppose σ_a and σ_b are the Thomson coefficients of the metals *A* and *B* respectively. If *dE is* the small resultant e.m.f. in the couple, then

$$dE = d\pi - (\sigma_a - \sigma_b)dT \qquad \text{... (1)}$$

Therefore, the thermo-electric power

$$\frac{dE}{dT} = \frac{d\pi}{dT} - (\sigma_a - \sigma_b) \qquad \text{... (1)}$$

Since Peltier and Thomson effects are reversible and Joule heating effect (which is irreversible) can be made negligible by considering that the e.m.f. and hence the current in the circuit is infinitely small, hence the net gain in entropy in the thermo-couple must be zero, *i.e.*

$$\Sigma \frac{\delta Q}{T} = 0 \qquad \text{... (2)}$$

We know that due to Peltier effect heat is absorbed at the hot junction and evolved at the cold junction while due to Thomson effect heat is evolved in *A* and absorbed in *B*. Now during the flowing of a charge *q* coulomb round thermo-couple, we have

Heat energy absorbed at the hot junction at temperature $T + dT$

$$= q\,(\pi + d\pi) \text{ joule}$$

Heat energy evolved at the cold junction at temperature

$$T = q\pi \text{ joule}$$

Heat energy evolved in A $= q.\sigma_a\, dT$

Heat energy absorbed in $B = q.\sigma_b\, dT$.

Now taking change in entropy due to a absorption of heat as positive and that due to evolution as negative, equation (2) gives

$$\frac{q(\pi + d\pi)}{T + dT} - \frac{q\pi}{T} - \frac{q\sigma_a dT}{T} + \frac{q\sigma_b dT}{T} = 0$$

or $$\left(\frac{\pi + d\pi}{T + dT} - \frac{\pi}{T}\right) - \left(\frac{\sigma_a dT}{T} - \frac{\sigma_b dT}{T}\right) = 0$$

or $$\frac{T.d\pi - \pi.dT}{T(T + dT)} - \frac{(\sigma_a - \sigma_b)dT}{T} = 0$$

Now dividing the numerator by dT, we have

$$\frac{T.\frac{d\pi}{dT} - \pi}{T(T + dT)} - \frac{(\sigma_a - \sigma_b)}{T} = 0 \qquad \text{... (3)}$$

Now proceeding to the limit $dT \to 0$, we have

$$T(T + dT) = T^2$$

so that equation (3) may be written as

$$\frac{T.\frac{d\pi}{dT} - \pi}{T^2} - \frac{(\sigma_a - \sigma_b)}{T} = 0 \qquad \text{... (4)}$$

This equation may be written as

$$\frac{d}{dT}\left(\frac{\pi}{T}\right) - \frac{(\sigma_a - \sigma_b)}{T} = 0 \qquad \text{... (5)}$$

This equation is called *Thermo-electric equation.* Equation (4) gives

$$\frac{d\pi}{dT}(\sigma_a - \sigma_b) = \frac{\pi}{T} \qquad \text{... (6)}$$

Substituting this value in equation (1), we get

$$\frac{dE}{dT} = \frac{\pi}{T}$$

$$\pi = T.\frac{dE}{dT} \qquad \text{... (7)}$$

Thus *the Peltier coefficient of a junction at absolute temperature T is the product of the absolute temperature and the thermo-electric power.* Differentiating equation (7), we get

$$\frac{d\pi}{dT} = \frac{dE}{dT} + T.\frac{d^2E}{dT^2}$$

or

$$\frac{d\pi}{dT} - \frac{dE}{dT} = T.\frac{d^2E}{dT^2} \quad \text{... (8)}$$

Also from equation (1),

$$\frac{d\pi}{dT} - \frac{dE}{dT}(\sigma_\alpha - \sigma_b) \quad \text{... (9)}$$

Comparing equations (8) and (9), we get

$$\sigma_a - \sigma_b = T.\frac{d^2E}{dT^2} \quad \text{... (10)}$$

This expression relates the Thomson coefficients with the absolute temperature and the first derivative of thermo-electric power.

Significance of Peltier Effect

In 1834 Jean Peltier (1785–1845) discovered a thermo-electric effect which is the converse of Seebeck effect. He found that when a current is passed through a thermo-couple whose two junctions are initially at the same temperature, then one of the junctions is heated, while the other is cooled; showing that heat energy is being liberated at one junction (thereby causing it to be heated) and absorbed at the other junction (thereby causing it to be cooled fig.) Thus by the passage of electric current through a thermo-couple with junctions initially at the same temperature, difference of temperature is developed between the two junctions.

The effect is known as Peltier effect. For a given pair of metals, the heating or cooling of the junction depends on the direction of the current. If a particular junction is heated by passing the current in one direction, the same is cooled when the direction of the current is reversed, *i.e.*, the *Peltier effect is reversible.*

It should be noted that the heating and the cooling of the two junctions in fig., caused by the Peltier effect, provides the conditions necessary for the Seebeck effect of fig. The direction of the resulting Seebeck current is opposite to the battery current. If this thermoelectric current did not oppose, but aided the battery current, it would result in the heated (cooled) junction being *still* farther heated (cooled), thereby resulting in a still greater current and so on, without limit.

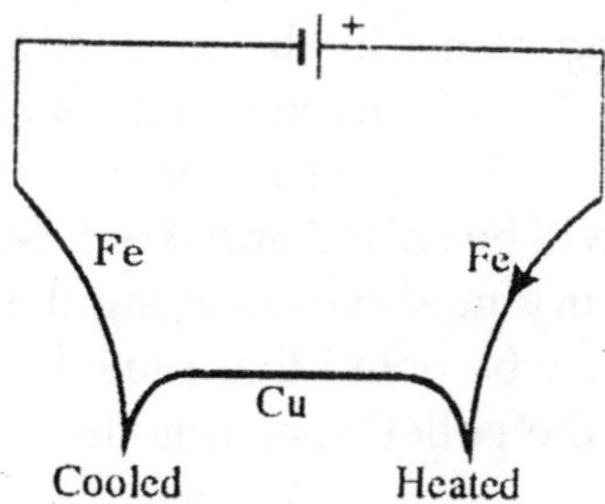

Experimental Demonstration of Peltier effect: The Peltier effect may be demonstrated by the following experiments:

Experiment: This experiment is due to Peltier himself. Two rods of antimony (Sb) and Bismuth (Bi) are joined as shown in fig. The thermo junctions *A* and *B* are placed in two glass bulbs joined by a thin tube containing a pellet of mercury. When no current is passed through the rods, junctions *A* and *B* are at the same temperature and the pellet of mercury lies symmetrically at the centre.

However, when a current is passed from a battery in a particular direction (say in the direction shown by arrow) through the couple, then there is evolution of heat at the junction A and absorption of heat at the other junction B. Therefore the junction A is heated and the junction *B* is cooled.

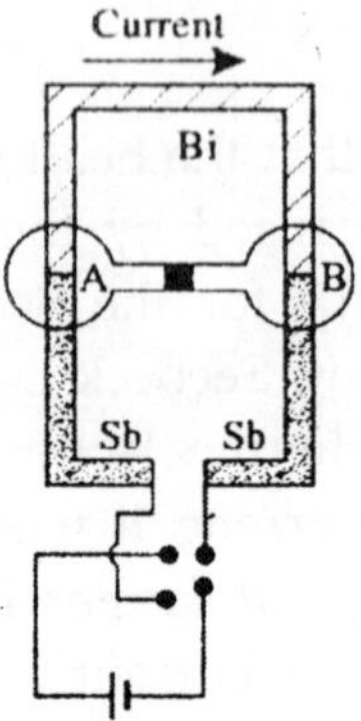

Due to heating of air in the bulb A and the cooling in the bulb *B*, the pellet of mercury moves along the direction $\overrightarrow{AB}$ (*i.e.*, from A to B). If now the direction of current is reversed, then A will be cooled and *B* will be heated. Therefore the pellet of mercury now moves along the direction $\overrightarrow{BA}$ (*i.e.* from *B* to A). It may be noted that Joule heating has no effect on the motion of the pellet since it is the same for the metals inside each bulb.

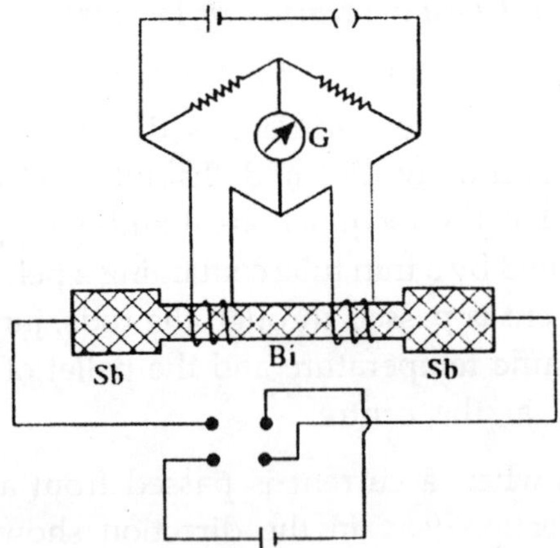

Experiment: S.G. Starling gave a better and easier method of demonstration of Peltier effect which is given below:

A compound bar of Bismuth with Antimony at the ends (as shown in fig.) is taken. Two identical coils consisting of insulated thin copper wire are closely wound at the junctions and are placed in the two arms of a Wheatstone bridge. The resistances in the other arms of the bridge are adjusted so that the bridge is initially balanced. When a current is passed through the bar, then due to Peltier effect, one junction is cooled, while the other is heated. Due to high temperature coefficient of copper, the resistance of the coil round the heated end increases; while the resistance of the coil round the cooled end decreases, therefore, the bridge is out of balance and the galvanometer shows a deflection.

If the deflection of the current is reversed, the effects are reversed and, therefore, the galvanometer now gives the deflection in opposite direction.

Explanation of Seebeck and Peltier Effects: According to electronic theory, the free electrons inside the metal move freely like the molecules of a gas in a container. Therefore, the free electrons are said to form a gas known as *electron gas*. This electron gas is uniformly distributed inside the entire volume of a metal. When the temperature of a metal is same everywhere, then the density and pressure of the electron gas inside the metal is also same everywhere.

The density and pressure of an electron gas differ from metal to metal even at the same temperature. When two metals are joined together, the electron gas diffuses from one metal to the other and vice versa in such a way that the net diffusion of the electron gas is from a metal at high pressure to that at low pressure.

Due to the diffusion of the electron gas, an e.m.f. is produced at the junction of the two metals : which opposes further diffusion of the electron gas. When this e.m.f. is sufficient to stop the further net diffusion of the electron gas, then the state of the dynamic equilibrium is reached. In the state of the dynamic equilibrium there exists a certain e.m.f at the junction of the two metals, which is known as Peltier e.m.f. p.

Seebeck Effect: When two dissimilar metals A and *B* (say Cu and Fe) are joined together to form a thermo-couple then at each junction of the thermo-couple, a Peltier e.m.f. is produced. If the two junctions are at the same temperatures, then the Peltier e.m.f.'s are equal and opposite.

Therefore, the net e.m.f. and hence the current in the thermo-couple is zero. If, however, the junctions are at different temperatures T_1 and T_2 ($T_2 > T_1$), the Peltier e.m.f.'s π_1 and π_2 at the junctions are no longer equal; so there is a resultant e.m.f. $\pi_2 - \pi_1$ in the cicruit, due to which a current flows in the thermo-couple. This explains Seebeck effect.

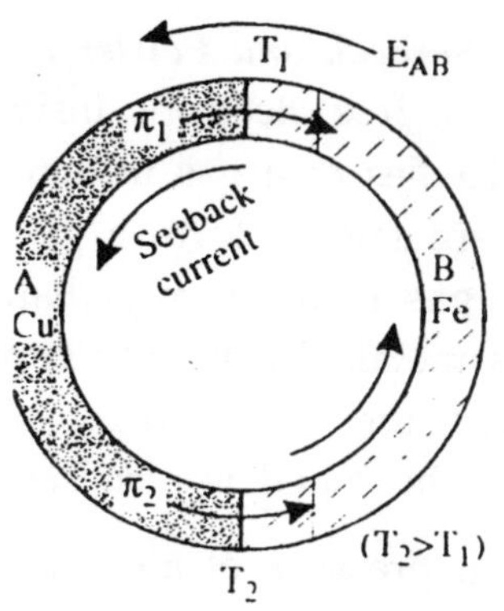

Peltier Effect: When two dissimilar metals A and *B* (say Cu and Fe) are joined together to form a thermo-couple then at each junction a Peltier e.m.f. is set up. Let the two junctions of the thermo-couple be at the same temperature, then the Peltier e.m.f.'s at the junctions are equal and opposite (*i.e.*, $\pi_1 = \pi_2 = \pi$ say). In the case of Fe-Cu junction, the Peltier e.m.f. acts from Cu to Fe. When a current flows through a junction from Fe to Cu by a battery of e.m.f. E_{AB}, it flows *against the Peltier e.m.f* and hence loses energy. This energy appears as heat and the junction is heated. At the other junction the current flows in the direction of the Peltier e.m.f. acts from Cu to Fe. When a current *flows against the Peltier e.m.f* and hence loses energy. This energy appears as heat and the junction is heated. At the junction the current flows in the direction of the Peltier e.m.f.; so that the Peltier e.m.f. itself does the work and energy is absorbed from the junction resulting into cooling of the junction. This explains Peltier effect.

Peltier Coefficient

The Peltier coefficient of a junction of two dissimilar metals is defined as the amount of energy absorbed or evolved (in joule) when unit charge (in coulomb) *flows* across the junction. It is denoted by π. Its value depends on the pair of metals in contact and on the temperature of the junction.

Thus, if a charge q coulomb passes across a junction having a Peltier coefficient n, then the energy evolved or absorbed at the junction = πq joule.

Now if e is the e.m.f. set up at the junction (Peltier e.m.f.) then the energy absorbed or evolved at the junction = eq joule.

Therefore $$\pi q = eq$$

or $$\pi = e.$$

Hence the Peltier coefficient expressed in joule per coulomb is numerically equal to the e.m.f. set up at the junction in volt.

Thomson Effect and its Prediction: Let us consider a thermo-couple made of two dissimilar metals A and B (say Cu and Fe) with its hot and cold junctions maintained at *absolute temperatures* T_2 and T_1 by a source and a sink of very large thermal capacity respectively. Let p_1 and p_2 be the Peltier e.m.f.'s at T_1 and T_2 K (kelvin) respectively, directed from the metal A to the metal B through the junctions. Since $T_2 > T_1$; $p_2 > p_1$: thus a resultant e.m.f. of $p_2 - p_1$ acts in the thermo-couple due to Peltier e.m.f. If we assume that the

Peltier e.m.f. is the only source of e.m.f. in the thermo-couple, then the thermo-electric e.m.f. developed in the thermo-couple is

$$e = \pi_2 - \pi_1. \quad \text{... (1)}$$

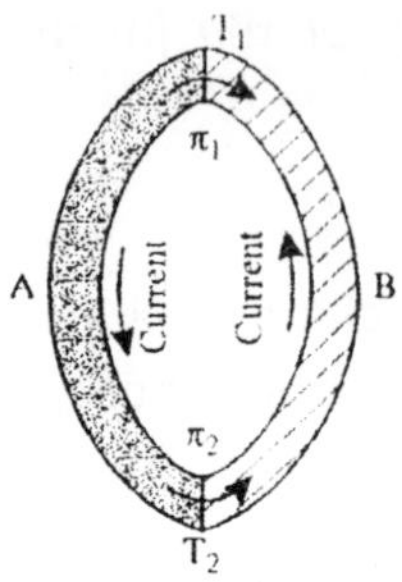

This e.m.f. gives rise to electric current flowing round the circuit in the anticlockwise direction as shown in fig.; which, due to Peltier effect, causes absorption of heat at the hot junction and evolution of heat at the cold junction. If q is the charge flowing in the thermo-couple in time t then during this time heat absorbed at the hot junction is $Q_2 = \pi_2 q$ and heat evolved at the cold junction is $Q_1 = \pi_1 q$.

The absorption of heat at the hot junction is made up by the source maintained at constant temperature T_2; hence Q_2 is the energy absorbed from the source. The heat evolved at the cold junction flows to the sink which maintains the junction at T_1K; hence heat given to the sink at T_1K *is* Q_1.

Since above processes (Seebeck and Peltier effects) are entirely reversible; for the current from an external source in the opposite direction and doing external work, heat may be absorbed from the junction at lower temperature T_1K and given to that at higher temperature T_2K; and Joule effect can be made negligible by considering a very small flow of current; therefore, the thermo-couple may be treated as a *reversible Carnot's engine* operating between the temperatures T_2K and T_1K of the hot and cold junctions. Now according to thermodynamics the net gain of entropy in a reversible cycle is zero, *i.e.*,

$$\Sigma \frac{Q}{T} = 0$$

or
$$\frac{Q_2}{T_2} - \frac{Q_1}{T_1} = 0$$

or
$$\frac{\pi_2 q}{T_2} - \frac{\pi_1 q}{T_1} = 0$$

or
$$\frac{\pi_2}{\pi_1} = \frac{T_2}{T_1}$$

or
$$\frac{\pi_2 - \pi_1}{\pi_1} = \frac{T_2 - T_1}{T_1}$$

or
$$(\pi_2 - \pi_1) = \frac{\pi_1}{T_1}(T_2 - T_1).$$

Substituting this value of $\pi_2 - \pi_1$ in equation (1), we get

Thermo-electric e.m.f., $e = \frac{\pi_1}{T_1}(T_2 - T_1).$ ... (2)

$\therefore$ $e \propto (T_2 - T_1)$ ... (3)

Thus, if the temperature T_1 of the cold junction is kept at constant temperature and T_2 is increased, then the resultant thermo-electric e.m.f. is directly proportional to the difference in temperature between hot and cold junctions, *i.e.*, the graph between e and $T_2 - T_1$ should be a straight line. But in actual experiments, it is seen that the relation between e and $T_2 - T_1$ is not linear; but *parabolic*.

This led Willian Thomson to predict that Peltier e.m.f. is not the only source of electro-motive force in the circuit and pointed out that there must be some other e.m.f. in the thermo-couple also. He pointed out that an additional source of e.m.f exists between different parts of the same metal due to the gradient of temperature throughout the metal. This effect is called Thomson effect.

Thomson Effect: When a temperature gradient is maintained between the different parts of the same metal, there exists a variation of potential along the metal, *i.e.,* e.m.f is developed in the metal due to the gradient of temperature. The effect is known as Thomson effect. As heat is either absorbed or evolved when current passes between two points having a difference of potential, therefore, the passage of electric current through such a metal (having temperature gradient) results in an absorption or evolution of heat in the body of the metal.

In case of a copper or antimony the potential increases from cold to hot end and decreases from hot to cold end; therefore, heat is absorbed when the current flows from the cold to the hot end, and is evolved when the current flows from the hot to the cold end. In the case of iron or bismuth the potential decreases from cold to the hot end and increases from hot to the cold end; therefore heat is evolved when the current flows from the cold to the hot end and is absorbed when the current flows from the hot to the cold end.

Thomson Coefficient: The Thomson coefficient σ of a metal is defined as the amount of heat energy absorbed or evolved when one coulomb charge flows in the metal between two points which differ in temperature by 1°C.

Thus if a charge q coulomb flows in a metal between two points having a temperature difference of 1°C, then

Heat energy absorbed or evolved $= \sigma q$ joule.

But if *E volt* be the Thomson e.m.f. developed between these points, then this energy must be equal to Eq joule.

Therefore $\sigma q = Eq$

or $\sigma = E.$

Hence Thomson coefficient of a metal is numerically equal to the e.m.f set up between two points differing in temperature by 1°C.

Thus the unit of Thomson coefficient is joule per coulomb per°C or volt per°C. The Thomson coefficient of a metal is positive when e.m.f is directed from lower to higher temperature as in copper and antimony and negative when e. m.f. is directed from higher to lower temperature as in iron and bismuth. Its value for lead is, however, practically zero. If σ is the Thomson coefficient, then the e.m.f, developed between the points at temperature T and $T + dT$ *is* $\sigma\, dT$. Therefore the resultant Thomson e.m.f. in a metal A whose ends are at temperatures T_1 and T_2 $(T_1 < T_2)$ is given by

$$\int_{T_1}^{T_2} \sigma dT$$

By analogy with specific heat, Thomson coefficient has been called the *specific heat of electricity* of the material.

Demonstration of Thomson Effect: An iron rod *ACB is* bent as shown in fig. and its ends *A* and *B* are immersed in mercury baths *A* and *B. Two* fine copper coils *P* and *Q* of equal resistance are wound over the two arms of the rod and packed round with asbestos wool and then connected in the two opposite gaps of a Wheatstone bridge.

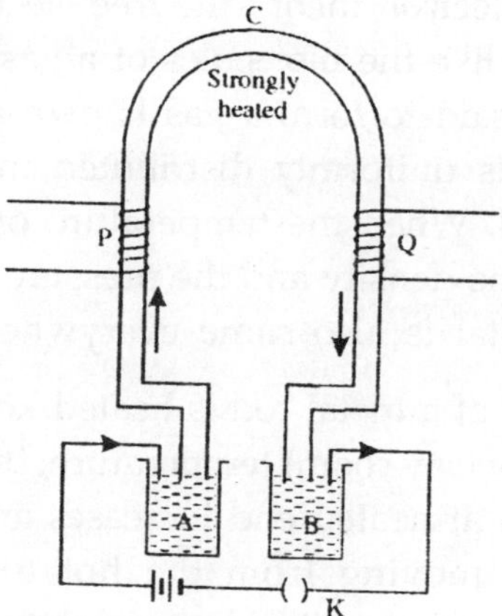

The end *C* of the rod is heated by Bunsen burner strongly while the ends *A* and *B* are kept cold in the mercury bath, so that there is a steep gradient of tempeature from C to A and C to B. Now a balance is obtained by varying the other two resistances R and S of the bridge.

Now a heavy current (say about 10 amp) is passed from A to B. It is found that the balance is disturbed and the galvanometer shows a deflection indicating that the resistance of P has increased while that of Q has decreased. It shows that heat is evolved in the arm AC in which the current flows from cold to hot end and absorbed in the arm CB in which current flows from hot to cold end. This experiment demonstrates the existence of Thomson effect in each of the two arms of the iron rod.

When the direction of the current is reversed, the galvanometer shows that the effects have also been reversed, thereby indicating that Thomson effect is reversible.

For a copper rod the effect is opposite and much smaller than that for iron rod.

For a lead rod no Thomson effect is observed.

Explanation of Thomson Effect

According to electron theory the free electrons inside the metal move freely like the molecules of a gas. Therefore, the free electrons are said to form a gas known as electron gas. The electron gas is uniformly distributed inside the entire volume of a metal. When the temperature of metal is same everywhere then the density and the pressure of the electron-gas inside the metal is also same everywhere.

When one end of a metal rod is heated keeping the other end at constant low (say room) temperature, then the pressure of the electron gas at heated end increases and so a number of electrons start moving from the hot to the cold end. Therefore the electron density increaes at the cold end and is more than that at the hot end.

Due to different electron densities in the hot and cold ends of the meal rod, an e.m.f. is developed which is sufficient to stop the further movement of the electrons from the hot to the cold end. Then the state of the dynamical equilibrium is reached. In the state of the dynamic equilibrium, there exists a certain e.m.f. between the ends of the rod which is known as Thomson e.m.f. The Thomson e.m.f. between two points of the rod differing in temperature by 1°C is called Thomson coefficient. It is now obvious that the Thomson e.m.f is due to the diffusion of electrons between differnet parts of the same metal which are at different temperatures.

In the case of copper or antimony rod, the Thomson e.m.f. acts from the part of the metal at lower temperature to the part at higher temperature. Therefore, when a current flows from the cold to the hot end, it flows in the direction of the Thomson e.m.f. and heat is absorbed from the rod. On the other hand, in the case of iron or antimony, the Thomson e.m.f. acts from higher temperature to lower temperature. Therefore when current flows from the cold to the hot end, it flows against the Thomson e.m.f. and hence heat is evolved.

E.M.F. in a Thermo-couple

Consider a thermo-couple made of two dissimilar metals *A* and *B* (say Fe and Cu) with its hot and cold junctions at absolute temperatures T_2 and $T_1 K$ respectively. If π_1 and π_2 are the Peltier coefficients at the cold and hot junctions respectively directed from metal *B* to metal *A* through the junctions; then the e.m.f. in the anticlockwise direction due to Peltier effect $= \pi_2 - \pi_1$. If σ_a and σ_b are the Thomson coefficients of the metals *A* and *B* respectively, then

Thomson e.m.f. in the metal A in anticlockwise direction,

$$E_A = -\int_{T_1}^{T_2} \sigma_\alpha dT$$

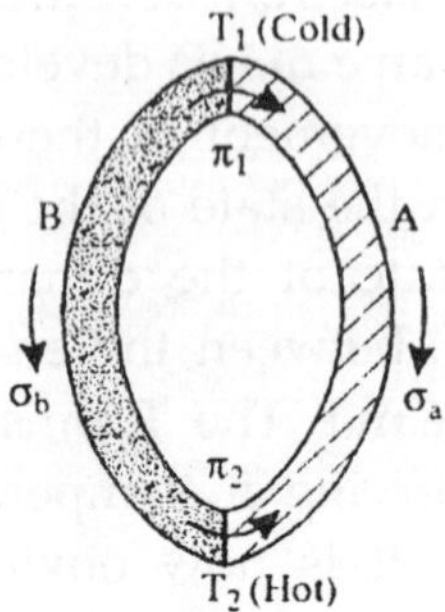

Thomson e.m.f. in the metal B in anticlockwise direction,

$$E_B = \int_{T_1}^{T_2} \sigma_b \, dT$$

Here σ_a and σ_b are both positive.

Hence the resultant e.m.f. in the circuit is

$$E_{AB} = \pi_2 - \pi_1 - \int_{T_1}^{T_2} \sigma_\alpha dT + \int_{T_1}^{T_2} \sigma_b dT$$

$$= \pi_2 - \pi_1 - \int_{T_1}^{T_2} (\sigma_\alpha - \sigma_b) dT \quad \text{... (1)}$$

In the case of Cu-Fe thermo-couple or antimony-bismuth thermo-couple σ_a is negative, therefore, for this couple

$$E_{AB} = \pi_2 - \pi_1 - \int_{T_1}^{T_2} (\sigma_a - \sigma_b) dT. \quad \text{... (2)}$$

Law of Intermediate Temperatures: Let us have three thermo-couples I, II, III made of the same dissimilar metals A and B. The temperatures of the junctions of first thermo-couple are T_1, and T_2, the temperatures of the junctions of second thermo-couple are T_1 and T_3 while the tempratures of the junctions of third thermo-couple are T, and T_3 as shown in fig. Let $T_3 > T_2 > T_1$. Then the e.m.f's in I, II and III thermo-couples are respectively.

$$[E_{AB}]_{T_1}^{T_2} = \pi_2 - \pi_1 - \int_{T_1}^{T_2} (\sigma_a - \sigma_b) dT \quad \text{... (1)}$$

$$\left[E_{AB}\right]_{T_2}^{T_3} = \pi_3 - \pi_2 - \int_{T_2}^{T_3} (\sigma_a - \sigma_b)\, dT \qquad \text{... (2)}$$

$$\left[E_{AB}\right]_{T_1}^{T_3} = \pi_3 - \pi_1 - \int_{T_1}^{T_3} (\sigma_a - \sigma_b)\, dT \qquad \text{... (3)}$$

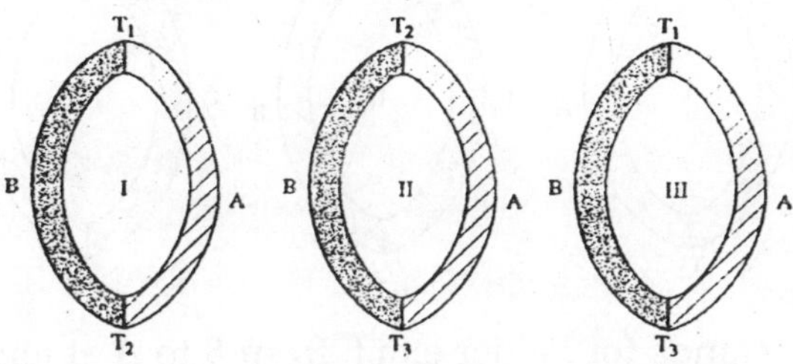

Adding equations (1) and (2), we get

$$\left[E_{AB}\right]_{T_1}^{T_2} + \left[E_{AB}\right]_{T_2}^{T_3} = \pi_3 - \pi_1 - \int_{T_1}^{T_3} (\sigma_a - \sigma_b)\, dT \qquad \text{... (4)}$$

Now comparing equations (3) and (4), we

$$\left[E_{AB}\right]_{T_1}^{T_2} + \left[E_{AB}\right]_{T_2}^{T_3} = \left[E_{AB}\right]_{T_1}^{T_3} \qquad \text{... (5)}$$

Thus the sum of e.m.f.'s of two couples formed of same metals A and B; but with their junctions at temperatures T_3, T_2 $(T_3 > T_2)$ and T_2, T_1, $(T_2 > T_1)$ respectively is equal to the e.m.f in a thermo-couple formed of the same metals with its junctions at temperatures T_3, T_1. This law is called the law of intermediate temperatures.

Law of Intermediate Metals

Let us have three thermo-coupples, I, II, III formed of metals *AB*, *BC* and *AC*; their hot and cold junctions being at temperatures T_2 and T_1K $(T_2 > T_1)$ respectively as shown in fig. Then the e.m.f.'s in the three couples are respectively

$$E_{AB} = \pi_{2BA} - \pi_{1BA} - \int_{T_1}^{T_2} (\sigma_a - \sigma_b)\, dT \qquad \text{... (1)}$$

$$E_{BC} = \pi_{2CB} - \pi_{1CB} - \int_{T_1}^{T_2} (\sigma_b - \sigma_c)\, dT \qquad \text{... (2)}$$

$$E_{AC} = \pi_{2CA} - \pi_{1CA} - \int_{T_1}^{T_2} (\sigma_a - \sigma_c)\, dT \qquad \dots (3)$$

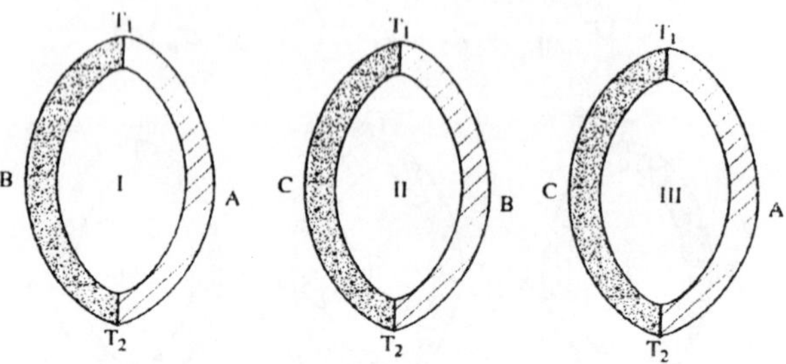

where π_{2BA} stands for Peltier e.m.f. from B to A at temperature T_2 for the thermo-couple AB, π_{1CB} stands for Peltier e.m.f. from C to B at temperature T_1 for the thermo-couple BC, etc.

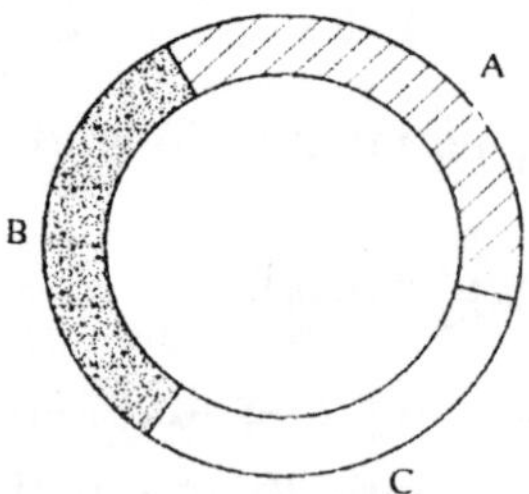

Now, if we arrange the three metals A, B, C in the form of a closed circuit and maintain its temperature T_1 everyhere, then the resultant e.m.f. in the circuit is zero. Moreover due to the absence of temperature gradient, Thomson e.m.f. is zero. Therefore the resultant e.m.f. in the circuit is only due to Peltier e.m.fs, so we must have

$$\pi_{1BA} + \pi_{1AC} + \pi_{1CB} = 0 \qquad \dots (4)$$

Similarly, if the temperature of the closed circuit is maintained at T_2K everywhere, then

$$\pi_{2BA} + \pi_{2AC} + \pi_{2CB} = 0 \qquad \dots (5)$$

Adding equations (1) and (2), we get

$$E_{AB} + E_{BC} = \pi_{2BA} + \pi_{2CB} - (\pi_{1BA} + \pi_{1CB})$$

$$-\int_{T_1}^{T_2}(\sigma_a - \sigma_C)dT \qquad \text{... (6)}$$

From equations (4) and (5)

$$\pi_{1BA} + \pi_{1CB} = -\pi_{1AC} + \pi_{1CA} \qquad (\text{since } \pi_{1AC} = -\pi_{1CA})$$

and

$$\pi_{2BA} + \pi_{2CB} = -\pi_{2AC} + \pi_{2CA} \qquad (\text{since } \pi_{2AC} = -\pi_{2CA})$$

Substituting these values in equation (6), we get

$$E_{AB} + E_{BC} = \pi_{2CA} + \pi_{1CA} - \int_{T_1}^{T_2}(\sigma_a - \sigma_C)dT \qquad \text{... (7)}$$

Comparing equations (3) and (7), we get

$$E_{AB} + E_{BC} = E_{AC}. \qquad \text{... (8)}$$

Thus for a given temperature difference between the hot and cold junctions, the sum of e.m.f's of the thermocouples AB and BC is equal to the e.m.f of the thermo-couple AC. This law is known as the law of intermediate metals.

Now it is obvious that if a third metal having same temperature at its ends is connected between a thermo-couple, then the e.m.f of the thermo-couple will remain unchanged, because the Thomson e.m.f. in the third metal is zero and Peltier e.m.f.'s at the junctions of the third metal are equal and opposite.

Importance of Seebeck Effect

In 1926 Thomas Johan Seebeck (1770–1831) discovered that *when two dissimilar metals are joined so as to form a closed circuit and a difference of temperature is established between junctions, an e.m.f is developed and hence electric current flows through the circuit.* The e.m.f. so proposed is called thermo-electric e.m.f. and the *phenomenon is known as Seebeck effect.* Such an arrangement of connecting two dissimilar metals is called *thermo-couple.* The magnitude of thermo-electric e.m.f. depends upon the nature of the two metals and on the temperature difference of their junctions.

Seebeak investigated the thermo-electric properties of a large number of metals and arranged them in a series known as the *thermo-electric series.* When a circuit is formed of the two metals of the series, the thermo-e.m.f. is greater, the farther the metals are apart in the series. Also the direction of the current, at the hot junction, is from the metal occuring earlier in series to the metal occuring later on it. The following is a selection from *Seebeck's series* → Bi, Ni, Co, Pt, Cu, Mn, Hg, Sn, Au, Ag, Zn, Cd, Fe, As, Sb, Te.

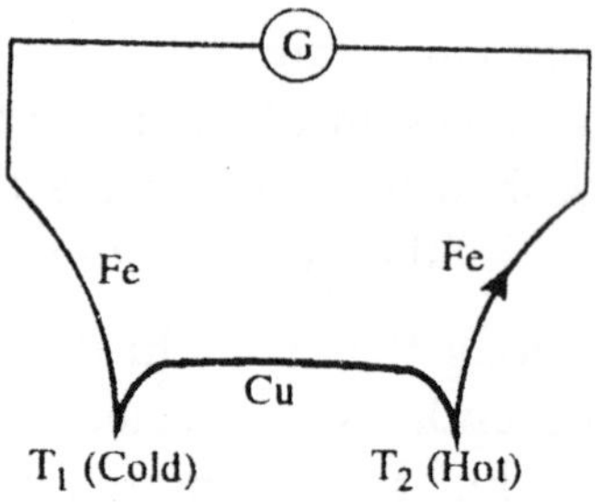

The position of a metal in this series, as Seebeck himself observed, is somewhat dependent on its purity. The thermo-couple of Cu and Fe is shown in Fig. The current in this couple flows from Cu to Fe through the hot junction. The thermo-e.m.f. for this couple is only 1.22 millivolt for a temperature difference $(T_2 - T_1)$ of 100° between their hot and cold junctions.

Variation of Thermo-e.m.f with Temperature: Let there be a thermo-couple of two dissimilar metals A and B. Let one junction (cold junction) be at 0°C and the temperature of the other junction (hot junction) be raised gradually. The thermo-e.m.f. in the circuit varies with the temperature of the hot junction and if thermo e.m.f. in the circuit is plotted against the temperature of the hot junction, the graph is parabola as shown in fig. The thermo-e.m.f. is zero when both the junctions are at the same temperature 0°C and gradually increases as the temperature of the hot junction increases.

At a particular temperature of the hot junction the thermo-e. m.f becomes maximum. This temperature of the junction (at which the e.m.f. in the thermo-couple is maximum) is called the *neutral temperature* T_n *for the thermo-couple.* The neutral temperature is constant for a given pair of metals forming the thermo-couple. If the temperature of the hot junction be further raised, the thermo-e.m.f. decreases and becomes zero at a particular temperature, called the temperature of inversion, T_i, Beyond the inversion temperature T_i, the thermo e.m.f. again increases; but in the reverse direction.

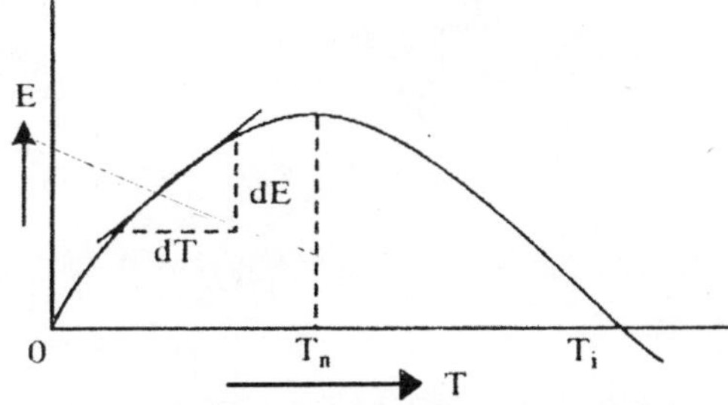

The temperature of inversion is as much above the neutral temperature as the temperature of the cold junction is below it. Therefore the inversion temperature is not constant for a given thermo-couple; but depends on the temperature of the cold junction. The temperature of inversion or the neutral temperature is obtained only when the metals of the thermocouple can remain without damage at these temperatures.

Statistical Power

The power of a statistical test is the probability that the test will reject a false null hypothesis (that it will not make a Type II error). As power increases, the chances of a Type II error decrease. The probability of a Type II error is referred to as the false negative rate (β). Therefore, power is equal to $1-\beta$.

Power analysis can either be done before (*a priori*) or after (*post hoc*) data is collected. *A* priori power analysis is conducted prior to the research study, and is typically used to determine an appropriate sample size to achieve adequate power. Post-hoc power analysis is conducted after a study has been completed, and uses the obtained sample size and effect size to determine what the power was in the study, assuming the effect size in the sample is equal to the effect size in the population.

Statistical tests use data from samples to determine if differences or similarities exist in a population. That is, do the criteria for selecting the samples divide the population into statistically distinct sub-populations. For example, to test the null hypothesis that the mean scores of men and women on a test do not differ, samples of men and women are drawn, the test is administered to them, and the mean score of one group is compared to that of the other group using a statistical test. The power of the test is the probability that the test will find a statistically significant difference between men and women, as a function of the size of the true difference between those two populations. Note that power is the probability of finding a difference that does exist, as opposed to the likelihood of declaring a difference that does not exist.

Statistical power depends on:

- The statistical significance criterion used in the test.
- The size of the difference or the strength of the similarity (that is, the effect size) in the population.
- The sensitivity of the data.

A significance criterion is a statement of how unlikely a result must be, if the null hypothesis is true, to be considered significant. The most commonly used criteria are probabilities of 0.05 (5 per cent, 1 in 20), 0.01 (1 per cent, 1 in 100), and 0.001 (0.1 per cent, 1 in 1000). If the criterion is 0.05, the probability of the difference must be less than 0.05, and so on. One way to increase the power of a test is to increase (that is, weaken) the significance level. This increases the chance of obtaining a statistically significant result (rejecting the null hypothesis) when the null hypothesis is false, that is, reduces the risk of a Type II error. But it also increases the risk of obtaining a statistically significant result when the null hypothesis is in fact true; that is, it increases the risk of a Type I error.

Calculating the power requires first specifying the effect size you want to detect. The greater the effect size, the greater the power.

Sensitivity can be increased by using statistical controls, by increasing the reliability of measures (as in psychometric reliability), and by increasing the size of the sample. Increasing sample size is the most commonly used method for increasing statistical power.

Although there are no formal standards for power, most researchers who assess the power of their tests use 0.80 as a standard for adequacy.

A common misconception by those new to statistical power is that power is a property of a study or experiment. In reality any statistical result that has a p-value has an associated power. For example, in the context of a single multiple regression, there will be a different level of statistical power associated with the overall r-square and for each of the regression coefficients. When determining an appropriate sample size for a planned study, it is important to consider that power will vary across the different hypotheses.

There are times when the recommendations of power analysis regarding sample size will be inadequate. Power analysis is appropriate when the concern is with the correct acceptance or rejection of a null hypothesis. In many contexts, the issue is less about determining if there is or is not a difference but rather with getting a more refined estimate of the population effect size. For example, if we were expecting a population correlation between intelligence and job performance of around 0.50, a sample size of 20 will give us approximately 80 per cent power (alpha = 0.05, two-tail).

However, in doing this study we are probably more interested in knowing whether the correlation is 0.30 or 0.60 or 0.50. In this context we would need a much larger sample size in order to reduce the confidence interval of our estimate to a range that is acceptable for our purposes. These and other considerations often result in the recommendation that when it comes to sample size, "More is better!"

Funding agencies, ethics boards and research review panels frequently request that a researcher perform a power analysis.

The argument is that if a study is inadequately powered, there is no point in completing the research.

The Probability

Probability is the likelihood or chance that something is the case or will happen. Probability theory is used extensively in areas such as statistics, mathematics, science and philosophy to draw conclusions about the likelihood of potential events and the underlying mechanics of complex systems.

Probability Distribution

In probability theory and statistics, a probability distribution identifies either the probability of each value of an unidentified random variable (when the variable is discrete), or the probability of the value falling within a particular interval (when the variable is continuous). The probability function describes the range of possible values that a random variable can attain and the probability that the value of the random variable is within any (measurable) subset of that range.

When the random variable takes values in the set of real numbers, the probability distribution is completely described by the cumulative distribution function, whose value at each real x is the probability that the random variable is smaller than or equal to x.

The concept of the probability distribution and the random variables which they describe underlies the mathematical discipline of probability theory, and the science of statistics. There is spread or variability in almost any value that can be measured in a population (e.g. height of people, durability of a metal, etc.); almost all measurements are made with some intrinsic error; in physics many processes are described probabilistically, from the kinetic properties of gases to the quantum mechanical description of fundamental particles. For these and many other reasons, simple numbers are often inadequate for describing a quantity, while probability distributions are often more appropriate.

There are various probability distributions that show up in various different applications. One of the more important ones is the normal distribution, which is also known as the *Gaussian distribution* or the bell curve and approximates many different naturally occurring distributions. The toss of a fair coin yields another familiar distribution, where the possible values are heads or tails, each with probability 1/2.

Random Variables

A random variable is a rigorously defined mathematical entity used mainly to describe chance and probability in a mathematical way. The structure of random variables was developed and formalised to simplify the analysis of games of chance, stochastic events, and the results of scientific experiments by retaining only the mathematical properties necessary to answer probabilistic questions. Further formalisations have firmly grounded the entity in the theoretical domains of mathematics by making use of measure theory.

Fortunately, the language and structure of random variables can be grasped at various levels of mathematical fluency. Set theory and calculus are fundamental.

Broadly, there are two types of random variables — discrete and continuous. Discrete random variables take on one of a set of specific values, each with some probability greater than zero. Continuous random variables can be realised with any of a range of values (e.g., a real number between zero and one), and so there are several ranges (e.g. zero to one half) that have a probability greater than zero of occurring.

A random variable has either an associated probability distribution (discrete random variable) or probability density function (continuous random variable).

Regression Analysis

In statistics, regression analysis is a collective name for techniques for the modelling and analysis of numerical data consisting of values of a dependent variable (response variable) and of one or more independent variables (explanatory variables).

The dependent variable in the regression equation is modelled as a function of the independent variables, corresponding parameters (constants), and an error term.

The error term is treated as a random variable. It represents unexplained variation in the dependent variable. The parameters are estimated so as to give a "best fit" of the data. Most commonly the best fit is evaluated by using the least squares method, but other criteria have also been used.

Regression can be used for prediction (including forecasting of time-series data), inference, hypothesis testing, and modelling of causal relationships. These uses of regression rely heavily on the underlying assumptions being satisfied. Regression analysis has been criticised as being misused for these purposes in many cases where the appropriate assumptions cannot be verified to hold. One factor contributing to the misuse of regression is that it can take considerably more skill to critique a model than to fit a model.

Sampling

Sampling is that part of statistical practice concerned with the selection of individual observations intended to yield some knowledge about a population of concern, especially for the purposes of statistical inference. Each observation measures one or more properties (weight, location, etc.) of an observable entity enumerated to distinguish objects or individuals. Survey weights often need to be applied to the data to adjust for the sample design. Results from probability theory and statistical theory are employed to guide practice.

The sampling process comprises several stages:

- Defining the population of concern.
- Specifying a sampling frame, a set of items or events possible to measure.
- Specifying a sampling method for selecting items or events from the frame.
- Determining the sample size.
- Implementing the sampling plan.
- Sampling and data collecting.
- Reviewing the sampling process.

Biased Samples

A biased sample is a statistical sample of a population in which some members of the population are less likely to be included than others. If the bias makes estimation of population parameters impossible, the sample is a non-probability sample.

An extreme form of biased sampling occurs when certain members of the population are totally excluded from the sample (that is, they have zero probability of being selected). For example, a survey of high school students to measure teenage use of illegal drugs will be a biased sample because it does not include home schooled students or dropouts. A sample is also biased if certain members are underrepresented or overrepresented relative to others in the population. For example, a "man on the street" interview which selects people who walk by a certain location is going to have an over-representation of healthy individuals who are more likely to be out of the home than individuals with a chronic illness.

Sampling Distribution

In statistics, a sampling distribution is the probability distribution, under repeated sampling of the population, of a given statistic (a numerical quantity calculated from the data values in a sample).

The formula for the sampling distribution depends on the distribution of the population, the statistic being considered, and the sample size used. A more precise formulation would speak of the distribution of the statistic as that for all possible samples of a given size, not just "under repeated sampling".

Simulation

Simulation is the imitation of some real thing, state of affairs, or process. The act of simulating something generally entails representing certain key characteristics or behaviours of a selected physical or abstract system.

However, the connection between simulation and dissembling later faded out and is now only of linguistic interest.

Simulation is used in many contexts, including the modelling of natural systems or human systems in order to gain insight into their functioning. Other contexts include simulation of technology for performance optimisation, safety engineering, testing, training and education. Simulation can be used to show the eventual real effects of alternative conditions and courses of action.

Key issues in simulation include acquisition of valid source information about the referent, selection of key characteristics and behaviours, the use of simplifying approximations and assumptions within the simulation, and fidelity and validity of the simulation outcomes.

Standard Deviation

In probability and statistics, the standard deviation is a measure of the dispersion of a collection of values. It can apply to a probability distribution, a random variable, a population or a data set. The standard deviation is usually denoted with the letter σ (lowercase sigma). It is defined as the root-mean-square (RMS) deviation of the values from their mean, or as the square root of the variance. Formulated by Galton in the late-1860s, the standard deviation remains the most common measure of statistical dispersion, measuring how widely spread the values in a data set are.

If many data points are close to the mean, then the standard deviation is small; if many data points are far from the mean, then the standard deviation is large. If all data values are equal, then the standard deviation is zero. A useful property of standard deviation is that, unlike variance, it is expressed in the same units as the data.

When only a sample of data from a population is available, the population standard deviation can be estimated by a modified standard deviation of the sample.

Statistical Graphics

Statistical graphics, also known as *graphical techniques*, are information graphics in the field of statistics used to visualise quantitative data.

Statistics and data analysis procedures can broadly be split into two parts: quantitative techniques and graphical techniques. Quantitative techniques are the set of statistical procedures that yield numeric or tabular output. Examples of quantitative techniques include hypothesis testing, analysis of variance, point estimation, confidence intervals, and least squares regression. These and similar techniques are all valuable and are mainstream in terms of classical analysis.

On the other hand, there is a large collection of statistical tools that we generally refer to as graphical techniques. These include: scatter plots, histograms, probability plots, residual plots, box plots, block plots and biplots. Exploratory data analysis (EDA) relies heavily on these and similar graphical techniques. Graphical procedures are not just tools used in an EDA context; such graphical tools are the shortest path to gaining insight into a data set in terms of testing assumptions, model selection and statistical model validation, estimator selection, relationship identification, factor effect determination, and outlier detection. In addition, good statistical graphics can provide a convincing means of communicating the underlying message that is present in the data to others.

Graphical statistical methods have four objectives:

- The exploration of the content of a data set.
- The use to find structure in data.
- Checking assumptions in statistical models.
- Communicate the results of an analysis.

If one is not using statistical graphics, then one is forfeiting insight into one or more aspects of the underlying structure of the data.

Statistical Survey

Statistical surveys are used to collect quantitative information about items in a population. Surveys of human populations and institutions are common in political polling and government, health, social science and marketing research. A survey may focus on opinions or factual information depending on its purpose, and many surveys involve administering questions to individuals. When the questions are administered by a researcher, the survey is called a *structured interview* or a *researcher-administered survey*. When the questions are administered by the respondent, the survey is referred to as a questionnaire or a self-administered survey.

Student's T-distribution

In probability and statistics, Student's *t*-distribution (or simply the *t*-distribution) is a probability distribution that arises in the problem of estimating the mean of a normally distributed population when the sample size is small. It is the basis of the popular Student's *t*-tests for the statistical significance of the difference between two sample means, and for confidence intervals for the difference between two population means. The Student's t-distribution is a special case of the generalised hyperbolic distribution.

The derivation of the *t*-distribution was first published in 1908 by William Sealy Gosset, while he worked at a Guinness Brewery in Dublin. He was prohibited from publishing under his own name, so the paper was written under the pseudonym Student. The *t*-test and the associated theory became well-known through the work of R.A. Fisher, who called the distribution "Student's distribution".

Student's distribution arises when (as in nearly all practical statistical work) the population standard deviation is unknown and has to be estimated from the data. Textbook problems treating the standard deviation as if it were known are of two kinds: 1) those in which the sample size is so large that one may treat a data-based estimate of the variance as if it were certain, and 2) those that illustrate mathematical reasoning, in which the problem of estimating the standard deviation is temporarily ignored because that is not the point that the author or instructor is then explaining.

Time Series

In statistics, signal processing, and many other fields, a time series is a sequence of data points, measured typically at successive times, spaced at (often uniform) time intervals. Time series analysis comprises methods that attempt to understand such time series, often either to understand the underlying context of the data points (Where did they come from? What generated them?), or to make forecasts (predictions). Time series forecasting is the use of a model to forecast future events based on known past events: to forecast future data points before they are measured. A standard example in econometrics is the opening price of a share of stock based on its past performance.

The term time series analysis is used to distinguish a problem, firstly from more ordinary data analysis problems (where there is no natural ordering of the context of individual observations), and secondly from spatial data analysis where there is a context that observations (often) relate to geographical locations. There are additional possibilities in the form of space-time models (often called *spatial-temporal analysis*). A time series model will generally reflect the fact that observations close together in time will be more closely related than observations further apart. In addition, time series models will often make use of the natural one-way ordering of time so that values in a series for a given time will be expressed as deriving in some way from past values, rather than from future values.

Methods for time series analyses are often divided into two classes: frequency-domain methods and time-domain methods. The former centre around spectral analysis and recently wavelet analysis, and can be regarded as model-free analyses well-suited to exploratory investigations. Time-domain methods have a model-free subset consisting of the examination of auto-correlation and cross-correlation analysis, but it is here that partly and fully-specified time series models make their appearance.

Variance

In probability theory and statistics, the variance of a random variable, probability distribution, or sample is one measure of statistical dispersion, averaging the squared distance of its possible values from the expected value (mean). Whereas the mean is a way to describe the location of a distribution, the variance is a way to capture its scale or degree of being spread out. The unit of variance is the square of the unit of the original variable. The positive square root of the variance, called the *standard deviation,* has the same units as the original variable and can be easier to interpret for this reason.

The variance of a real-valued random variable is its second central moment, and it also happens to be its second cumulant. Just as some distributions do not have a mean, some do not have a variance. The mean exists whenever the variance exists, but not vice versa.

Mechanical Statistics

In Physical Chemistry, the study of statistical mechanics can be divided into two main classes.

1. Classical Statistics or Maxwell–Boltzmann Statistics.
2. Quantum Statistics.

Importance of Degeneracy

Case (i)

Weak Degeneracy: At $T > T_F$ (*i.e.*, at intermediate temperatures) the Fermi gas is said to be slightly degenerate. In this case $kT > \varepsilon_F(\dot{0})$, then ε_F is negative or α is positive and $A < 1$.

For A < 1, we can write

$$\frac{1}{\frac{1}{A}e^x + 1} = \left\{\frac{1}{A}e^x + 1\right\} = \left(Ae^{-x}\right)\left(1 + Ae^{-x}\right)^{-1}$$

$$= Ae^{-x}\left(1 - Ae^{-x} + A^{-2x} - \ldots\right).$$

Therefore $$f_1(\alpha) = \frac{2}{\sqrt{\pi}}\int_0^\infty \frac{x^{1/2}dx}{\frac{1}{A}e^x + 1}$$

$$= \frac{2}{\sqrt{\pi}}\left[A\int_0^\infty e^{-x}x^{1/2}dx - A^2\int_0^\infty e^{-2x}x^{1/2}dx + A^3\int_0^\infty e^{-3x}x^{1/2}dx - ...\right]$$

$$= A - \frac{A^2}{2^{3/2}} + \frac{A^3}{3^{3/2}} - ...... \quad ...(1)$$

Similarly, $$f_2(\alpha) = A - \frac{A^2}{2^{5/2}} + \frac{A^3}{3^{5/2}} - ...... \quad ...(2)$$

Using these values of f_1, (α) and f_2 (α), the new equations take the form

$$n = g_s.\frac{V}{h^3}(2\pi mkT)^{3/2}\left[A - \frac{A^2}{2^{3/2}} + \frac{A^3}{3^{3/2}} - ......\right] \quad ...(3)$$

$$E = \frac{3}{2}g_s.\frac{V}{h^3}(2\pi mkT)^{3/2}.kT\left[A - \frac{A^2}{2^{5/2}} + \frac{A^3}{3^{5/2}} - ......\right] \quad ...(4)$$

Dividing equation (4) by (3), we get

$$\frac{E}{n} = \frac{3}{2}kT\left[A - \frac{A^2}{2^{5/2}} + \frac{A^3}{3^{5/2}} - ......\right]\left[A - \frac{A^2}{2^{3/2}} + \frac{A^3}{3^{3/2}} - ......\right]^{-1}$$

i.e. $$E = \frac{3}{2}nkT\left[1 + \frac{A}{2^{5/2}} + \frac{A^2}{3^{5/2}} - ......\right] \quad ...(5)$$

To the first approximation we can write

$$n = g_s.\frac{V}{h^3}(2\pi m\, kT)^{3/2}A$$

$$E = \frac{3}{2}g_s.\frac{V}{h^3}(2\pi m\, kT)^{3/2}.kTA$$

$$\frac{E}{n} = \frac{3}{2}kT \text{ or } E = \frac{3}{2}nkT. \quad ...(6)$$

which is well known relation for a perfect gas in classical statistics.

A comparison of equations (5) and (6) shows that ideal Fermi-Dirac gas deviates from perfect gas behaviour and this deviation, as we know, is called degeneracy. It is obvious that degeneracy is the function *of* A (or $e^{-\alpha}$). Greater is the value of A, more marked will be the degeneracy. Hence for $A < 1$ or $T > T_F$, the Fermi-gas is slightly degenerate.

Case (ii)

Strong Degeneracy: When α is large and negative, $A = e^{-\alpha} >> 1$. As degeneration increases with increase of A, therefore in this case degeneracy becomes more prominent. Further to the first approximation from eqn. (3), we have

$$A \approx \frac{1}{g_s} \cdot \frac{n}{V} \cdot \frac{h^3}{(2\pi mkT)^{3/2}} \qquad \text{... (7)}$$

This eqn. shows that the gas will be strongly degenerate at low temperatures and high particle densities $\frac{n}{V}$. The evaluation of integrals f_1 (α) and f_2 (α) under these conditions is complicated.

We shall discuss this case *of* strong degeneracy at absolute zero, *i.e.* when $T = 0$. When $T \to 0$, $A \to \infty$. In this case the Fermi-Dirac gas is completely degenerate.

At $T = 0$, we have

$$f(\varepsilon) = \frac{1}{\frac{1}{A} e^{\varepsilon/kT} + 1} = \frac{1}{e^{(\varepsilon - \varepsilon_F)/kT} + 1} = 1 \text{ for } 0 \le \varepsilon \le \varepsilon_F(0)$$

$$= 0 \text{ for } \varepsilon > \varepsilon_F(0)$$

so that

$$dn = g_s \cdot \left(\frac{4\pi mV}{h^3}\right) \cdot (2m)^{1/2} \varepsilon^{1/2} \, d\varepsilon \text{ for } 0 \le \varepsilon \le \varepsilon_F(0)$$

$$= 0 \text{ for } \varepsilon > \varepsilon_F(0)$$

where ε_F (0) *is* given

Now the total internal energy of perfect Fermi-Dirac gas at T = 0 i.e. zero point energy of Fermi gas is

$$E_0 = \int_0^{\varepsilon_F(0)} \varepsilon dn = g_s \cdot \left(\frac{4\pi mV}{h^3}\right)(2m)^{1/2} \int_0^{\varepsilon_F(0)} \varepsilon^{3/2} d\varepsilon$$

$$= g_s \left(\frac{4\pi mV}{h^3}\right)(2m)^{1/2} \cdot \left[\frac{\varepsilon^{5/2}}{5/2}\right]_0^{\varepsilon_F(0)}$$

$$= \frac{2}{5} g_s \left(\frac{4\pi mV}{h^3}\right)(2m)^{1/2} \cdot [\varepsilon_F(0)]^{5/2}.$$

Using above equation gives

$$E_0 = \frac{3nh^2}{10m}\left[\frac{3n}{4\pi V g_s}\right]^{2/3} = \frac{3}{5} n\varepsilon_F(0). \qquad \text{... (8)}$$

Now the pressure at $T = 0$ (i.e. zero point pressure) is given by

$$P_0 = \frac{2E_0}{3V}\left[\text{Since } P = -\left(\frac{\partial E}{\partial T}\right)_{T,S} = \frac{3E}{3V}\right]$$

$$= \frac{1}{5} \cdot \frac{nh^2}{Vm}\left(\frac{3n}{4\pi g_s V}\right)^{2/3}. \qquad \text{... (9)}$$

From equations (8) and (9) it is obvious that a strongly degenerate Fermi-Dirac gas possesses energy and exerts a pressure even at 0*K*, quite unlike Bose–Einstein and classical gases where the energy and pressure at absolute zero are zero.

Electron Gas in Metals

A metal can be considered to be composed of a system of fixed positive nuclei and a number of mobile electrons. These mobile electrons are assumed to move freely in the metal like the particles of a gas and constitute a perfect gas known as electron gas.

On this assumption the classical statistics could explain to a certain extent the various properties of metals dependent on the motion of free electrons in metals such as electrical and thermal conductivities, thermoelectricity, thermionic emission, magnetic properties of metals and photoelectric effect, etc. But in certain cases, the chief among them being the specific heat of metals, very serious difficulties were encountered in the use of classical statistics. That is why the theory of the electron gas was discredited to some extent.

Sommerfeld, in *1928*, however, revived the electron theory of metals on the basis of the new quantum statistics. According to him, the electrons in metals are not completely free but only partially so, in the sense that though they are not bound to any particular atomic system, yet they are bound to the metal as a whole. Therefore, the interior of the metal is to be conceived as a region of uniform potential, positive relative to free space, so that work is required to be done to extract an electron from the metal.

The electrons in metals cannot, therefore, be compared to the free particles of a gas obeying the classical statistics. Moreover due to their light mass and dense packing, the electrons in the metals should be assimilated to the particles of a gas under very high compression, hence to a degenerate gas. Further, since these electrons are assumed to obey Pauli's exclusion principle, they should obey the Fermi-Dirac statistics.

For electron $s = \frac{1}{2}$ so that $g_s = 2s + 1 = 2$.

The Fermi energy at 0K is

$$\varepsilon_F = \frac{h^2}{2m}\left[\frac{3n}{4\pi V g_s}\right]^{2/3} = \frac{h^2}{2m}\left[\frac{3n}{4\pi V.2}\right]^{2/3}.$$

$$= \frac{h^2}{8m}\left(\frac{3n}{\pi V}\right)^{2/3} = 0.625\times10^{-17}\rho^{2/3} \text{ joule or } 39\rho^{2/3}\text{ eV} \quad \text{... (1)}$$

where $\rho = \left(\frac{mn}{V}\right)^{2/3}$ kg/m³ *is* the density of the electron gas. For conduction electrons in metals $\rho \approx 0.1$ kg/m³.

The Fermi-temperature T_F for electron gas is

$$T^F = \frac{\varepsilon_F(0)}{k} = \frac{h^2}{8mk}\left(\frac{3n}{\pi V}\right)^{2/3} = \left(4.52 \times 10^5 \rho^{3/2}\right)K \qquad \text{... (2)}$$

Thus electron gas below 10^5 K temperature is degenerate.

The degeneracy factor of an electron gas, from eqn. (7), is

$$A = \frac{1}{g_s}.\frac{n}{v}.\frac{h^3}{(2\pi mkT)^{3/2}} = \frac{1}{2}.\frac{n}{V}.\frac{h^3}{(2\pi mkT)^{3/2}}$$

As $T \to 0. A \to \infty$, therefore $\dfrac{1}{e^{\alpha+x}+x} = \dfrac{1}{\frac{1}{A}e^{\alpha x}+x} \to 1.$

and that for low temperatures

$$n = g_s.\frac{V}{h^3}(2\pi mkT)^{3/2}.\frac{1}{\sqrt{\pi}}\int_0^\infty x^{1/2}dx$$

$$= \frac{4V}{\sqrt{\pi}}\frac{(2\pi mkT)^{3/2}}{h^3}\int_0^A x^{1/2}dx.$$

Here we have put $g_s = 2$ and have replaced the upper limit by A at low temperatures near absolute zero since $A \to \infty$, when $T \to 0$

or $$n = \frac{4V}{\sqrt{\pi}}\frac{(2\pi mkT)^{3/2}}{h^3}.\frac{A^{3/2}}{\frac{3}{2}}.$$

i.e. $$A = \frac{h^2}{2\pi mkT}.\left(\frac{3n}{8\pi V}\right)^{2/3} = \frac{h^2}{2m^{5/2}.kT}\left(\frac{3\rho}{8\pi}\right)^{2/3} \qquad \text{.... (3)}$$

This equation represents the degeneracy factor of electron gas at low temperatures near absolute zero.

Now, substituting $h = 6.6 \times 10^{-34}$ joule-sec, $m = 9 \times 10^{-31}$ kg., $k = 1.38 \times 10^{-23}$ joule/kelvin and $\rho = 0.1$ kg/m^3.

$$A = \frac{4.56 \times 10^5}{T}$$

This means than at low temperature the electron gas is strongly degenerate.

Zero point energy of the electron gas, from eqn. (8) is

$$E_0 = \frac{3nh^2}{10m}\left(\frac{3n}{4\pi V g_s}\right)^{2/3} = \frac{3nh^2}{10m}.\left(\frac{3n}{4\pi V.2}\right)^{2/3}$$

$$= \frac{3nh^2}{40m}\left(\frac{3n}{\pi V}\right)^{2/3} = \frac{3}{5}n\varepsilon_F(0). \qquad \text{... (4)}$$

Zero point pressure of the electron gas, from eqn. (2), is

$$P_0 = \frac{1}{5}.\frac{n}{V}.\frac{h^2}{m}\left(\frac{3n}{4\pi g_s V}\right)^{2/3} = \frac{1}{5}.\frac{n}{V}.\frac{h^2}{m}\left(\frac{3n}{8\pi V}\right)^{2/3}$$

$$= \frac{nh^2}{20mV}\left(\frac{3n}{\pi V}\right)^{2/3}. \qquad \text{... (5)}$$

At normal temperature the pressure of the electron gas comes out to be sufficiently high of the order of $\approx 10^5$ atmosphere.

The electronic contribution to the specific heat of metals at low temperatures is given by

$$C_V = \left(\frac{\partial E}{\partial T}\right)_V = \frac{\partial}{\partial T}\left[\frac{3}{5}n\varepsilon_F(0)\left\{1 \div \frac{5}{12}\left(\frac{\pi k T^2}{\varepsilon_F(0)}\right)\right\}\right]_V$$

$$= \frac{1}{2}nk\pi^2\left(\frac{\pi k T^2}{\varepsilon_F(0)}\right) \qquad \text{... (6)}$$

$$= \lambda T$$

where $\lambda = \dfrac{nk^2\pi^2}{2\varepsilon_F(0)}$ is constant quantity, independent of temperature. *i.e.* $C_v \propto T$

Thus the electronic contribution to the specific heat is proportional to the absolute temperature and vanishes at the absolute zero.

PROBLEMS

1. *Calculate the Fermi energy in electron volt for sodium assuming that it has one free electron per atom. Given density of sodium = 0.97 g cm ³, atomic weight of sodium is 23.*

Solution: The Fermi-energy is given by

$$\varepsilon_F(0) = \frac{h^2}{8m}\left(\frac{3n}{\pi V}\right)^{2/3}.$$

Assuming one electron per sodium atom, the electron density $\frac{n}{V}$ is given by

$$\frac{n}{V} = \frac{N_0 \rho}{W}$$

Where

Avogadro Number $N_0 = 6 \times 10^{26}$ atoms/kg. mole.

Density of sodium, $\rho = 0.97$ g $cm^{-3} = 0.97 \times 10^3$ kg./m^3

Atomic weight of sodium, $W = 23$

$\therefore$ Electron density,

$$\frac{n}{V} = \frac{\left(6\times10^{26}\ \text{atoms/kg.mole}\right)\left(0.97\times10^{3}\ \text{kg./m}^3\right)}{23}$$

$$= 2.53 \times 10^{28}\ \text{electrons/m}^3.$$

Also Planck's constant, $h = 6.62 \times 10^{-34}$ joule–sec. Mass of the electron, $m = 9.1 \times 10^{-31}$ kg. Therefore the Fermi energy ε_F (0) is given by

$$\varepsilon_F(0) = \frac{h^2}{8m}\left(\frac{3}{\pi}\cdot\frac{n}{V}\right)^{2/3}$$

$$= \frac{\left(6.62\times10^{-34}\ \text{joule-sec}\right)^2}{\left(8\times9\times10^{-31}\ \text{kg.}\right)} \times \left[\frac{3}{3.14} 2.53\times10^{18}\ electrons/m^3\right]^{2/3}$$

$$= 5.032 \times 10^{-19} \text{ joule}$$

$$= \frac{5.032 \times 10^{-19}}{1.6 \times 10^{-19}} eV = 3.145 \text{ eV}.$$

2. *Show that average energy at 0°K will be $\frac{3}{5}$ times of Fermi energy.*

Solution: The number of electrons in an electron gas having energies between ε and ε + *d*ε is given by

$$n(\varepsilon)d\varepsilon = \frac{8\pi mV}{h^3}\sqrt{(2m)}.\frac{\varepsilon^{1/2}d\varepsilon}{e^{(\varepsilon-\varepsilon_F)/kT}+1} \quad \text{... (1)}$$

where *m is* mass of electron, the Fermi energy ε_F is

$$\varepsilon_F(0) = \frac{h^2}{8m}\left(\frac{3N}{\pi V}\right)^{2/3}$$

$$\therefore \quad \frac{V}{h^3} = \frac{3N}{8\pi m 2\sqrt{(2m)}}\left[\varepsilon_F(0)\right]^{-3/2}$$

Substituting this in equation (1), we get

$$n(\varepsilon)d\varepsilon = \frac{3}{2}N\left[\varepsilon_F(0)\right]^{-3/2}\frac{\varepsilon^{1/2}d\varepsilon}{e^{(\varepsilon-\varepsilon_F)/kT}+1}$$

At $T = 0$, all of the electrons have energies less than or equal to ε_F (0) (*i.e.* $\varepsilon \le \varepsilon_F$) so that at $T = 0$, we have

$$e(\varepsilon^{-\varepsilon F)/kT} = e^{-\infty} = 0.$$

∴ At absolute zero

$$n(\varepsilon)d\varepsilon = \frac{3N}{2}\left[\varepsilon_F(0)\right]^{-3/2}\varepsilon^{1/2}d\varepsilon \quad \text{... (2)}$$

Total energy at absolute zero

$$E_0 = \int_0^{\varepsilon_F(0)} \varepsilon n(\varepsilon)d\varepsilon$$

$$= \frac{3}{2}N\left[\varepsilon_F(0)\right]^{-3/2}\int_0^{\varepsilon_F(0)} \varepsilon^{3/2}d\varepsilon$$

$$= \frac{3}{5} N\varepsilon_F(0)$$

$$E_0 = \int_0^{\varepsilon_F(0)} \varepsilon n(\varepsilon) d\varepsilon$$

$$= \frac{3}{2} N[\varepsilon_F(0)]^{-3/2} \int_0^{\varepsilon_F(0)} \varepsilon^{3/2} d\varepsilon$$

$$= \frac{3}{5} N\varepsilon_F(0)$$

Therefore, average energy at absolute zero

$$\overline{\varepsilon_0} = \frac{E_0}{N} = \frac{3}{5}\varepsilon_F(0)$$

Specific Heat Anomaly of Metals and its Solution: Experiments show that the specific heat of metals referred to 1 gram atom does not exceed a value of $3R$ even at very large temperatures.

Theoretically, the specific heat of metals shall get contribution from the atom as well as the free electrons, *i.e.*

C_V (metals) = C_V (atomic) + C_V (electronic).

According to the law of equipartition of energy of classical statistics, the average energy associated with each quadratic term is $\frac{1}{2}kT$. Further, according to the classical kinetic theory of matter, the atoms of a solid substance are at rest under the action of their mutual attractions and repulsions at the absolute zero of temperature. The energy of solids in this state is assumed to be zero. When the temperature is raised, the atoms are set into vibrations about their positions of equilibrium and the restoring force proportional to displacement is produced as long as the amplitude of vibration is not too large. Thus the vibrations of the atoms are simple harmonic in nature.

In S.H.M., the energy of the atom along one coordinate can be represented as the sum of two quadratic terms, *viz.*,

$$\frac{1}{2}m\dot{x}^2 + \frac{1}{2}Kx^2.$$

where K is the force constant, x is the displacement of the atom from mean position and $\dot{x} = \frac{dx}{dt}$. Here the first term represents the kinetic energy and the second term represents the potential energy.

According to law of equipartition of energy, the average energy associated with the motion of an atom along one coordinate

$$= 2 \times \frac{1}{2}kT = kT.$$

As each atom is free to vibrate along three coordinate axes, the total average energy of each atom = $3kT$.

The total energy of a gram atom of solid containing N (= Avogadro number) atoms is given by

$$E_{atomic} = 3NkT = 3RT$$

where R = Nk gas constant for a gram atom.

Therefore, atomic contribution to specific heat referred to 1 gram atom of metal is

$$C_V\,(\text{atomic}) = \left[\frac{\partial E_{atomic}}{\partial T}\right]_V = \frac{\partial}{\partial T}(3RT) = 3R. \qquad ...(1)$$

As the electrons inside the metal are assumed to be free to move along three axes, they do not possess the potential energy; but possess only the kinetic energy. Accor .ing to law of equipartition of energy, the average energy associated with the motion

$$= 3.\frac{1}{2}kT = \frac{3}{2}kT.$$

Assuming one free electron per atom in the metal, the total energy of a gram atom of solid containing N atoms is

$$E_{\text{electronic}} = \frac{3}{2} NkT = \frac{3}{2} RT.$$

Therefore, the electronic contribution to specific heat referred to 1 gram atom of metal is

$$C_V\,(\text{electronic}) = \left[\frac{\partial E_{\text{electronic}}}{\partial T}\right]_V = \frac{\partial}{\partial T}\left(\frac{3}{2}R\right)T = \frac{3}{2}R \qquad \text{... (2)}$$

Hence classically the specific heat referred to 1 gram atom of metal is

$$C_V = 3R + \frac{3}{2}R = \frac{9}{2}R. \qquad \text{... (3)}$$

which is too high as compared to the experimental value *3R*. *This* specific heat anomaly of metals is the chief problem which the classical statistics could not solve. This specific heat anomaly was solved by Fermi-Dirac statistics in a very elegant manner.

According to Fermi-Dirac statistics the electronic contribution to the specific heat per gram atom of metals at low temperature is

$$C_V\,(\text{electronic}) = \frac{1}{2}.\frac{Nk^2\pi^2}{\varepsilon_F(0)}T = \lambda T \qquad \text{... (4)}$$

where $$\lambda = \frac{1}{2}\cdot\frac{Nk^2\pi^2}{\varepsilon_F(0)} \qquad \text{... (5)}$$

Hence N is Avogadro Number.

Obviously, the electronic contribution to the specific heat of a metal at low temperatures is proportional to the temperature and vanishes at absolute zero in agreement with the experimental observations.

On substituting numerical values in the above relation it is found that the contribution of the electrons to the specific heat at ordinary temperatures is about 1 per cent of heat of the atoms.

For example, for silver C_V (electronic) = 0.046 cal per gram atom at 300K, which is only 0.7 per cent of the ordinary specific heat of silver at 300K, and hardly adds any thing to it.

Hence we should not expect to detect the electronic specific heat in specific heat measurements at ordinary temperatures. It is only when very high temperatures of the order of 10,000K are reached, that the tight packing of the electrons gradually becomes loosened and the electrons make a noticeable contribution to the specific heat.

According to Debye's theory the specific heat at extremely low temperatures is found to be proportional to T^3. Hence the specific heat of metals at low temperatures can be expressed as

$$C_V = \beta T^3 + \lambda T. \qquad \text{... (6)}$$

where the first term refers to the atomic contribution to the specific heat according to Debye's theory and the second term refers to the electronic contribution to the specific heat according to Fermi-Dirac statistics. At very low temperatures the atomic specific heat becomes small and the electronic specific heat becomes relatively high. This has been verified by many scientists for a variety of metals.

For example, according to eqn. (6), the atomic specific heat for silver becomes small compared to its electronic specific heat from the temperature range 3K.

In specific heat measurements Keesom and Cok found that in the range from 1.5K to 3K the specific heat varies according to eqn. (4): but above 3K, the T^3 term predominates. This suggests that below 3K for silver, free electrons become the chief contributor to the specific heat. Thus the Fermi-Dirac statistics has solved problem of specific heat of metals.

Significance of Quantum Statistics

The Classical Statistics is so called because it requires the classical results of Maxwell-Boltzmann velocity distribution of particles of an assembly in equilibrium in interpreting the statistical system of interest. Classical Statistics interpreted successfully many ordinarily observed phenomena such as temperature, pressure, energy, etc. But it failed to account adequately for several other experimentally observed phenomena such as black body radiation, photoelectric effect, specific heat at low temperatures, etc. This failure of Classical Statistics forced the issue in favour of the new quantum idea of discrete exchange of energy between systems and alongwith it a new statistics, known as quantum statistics, was investigated.

The quantum statistics was first formulated in 1924 by Bose in the deduction of Planck's radiation law by purely statistical reasoning on the basis of certain fundamental assumptions radically different from those of classical statistics. Einstein in the same year utilised practically the same principles in evolving the kinetic theory of gases, as a substitute for the classical Boltzmann Statistics. Thus a new quantum statistics, known as Bose-Einstein statistics, came to be accepted. Two years later in 1926, Fermi and Dirac quite independently modified Bose-Einstein statistics in certain cases, on the basis of additional principle, suggested first by Pauli in connection with electronic structure of atoms and known as Pauli's exclusion principle according to which in any given orbit in an atom there can be one and only one electron, it being impossible for two electrons to coexist in the same quantum state.

In statistical mechanics this exclusion principle takes the form that two or more phase points cannot possibly occupy the same phase cell. This led to the recognition of a second kind of quantum statistics, called, the Fermi-Dirac statistics. Thus quantum statistics can again be put into two sub-classes :

1. *Base-Einstein Statistics:* This is applicable to the identical, indistinguishable particles of zero or integral spin. These particles are called Bosons. The examples of Bosons are helium atoms at low temperature and the photons.
2. *Fermi-Dirac Statistics:* This is applicable to the identical, indistinguishable particles of half *spin.* These particles obey Pauli's exclusion principle and are called Fermions or Fermi particles. The examples of Fermions are electrons, photons, neutrons, etc.

The essential difference between the three statistics may be illustrated in the following simple manner:

Let there be only two points of a collection and only two cells to be occupied.

In *classical statistics* each particle has a recognisable individuality, hence the particles are distinguishable and each particle is as likely to be in one cell as in the other. This means that in classical statistics, each or both of the particles can occupy any one of the two cells. Thus in this case we have four possible arrangements, each of which is counted in assigning a probability to the distribution (refer fig.).

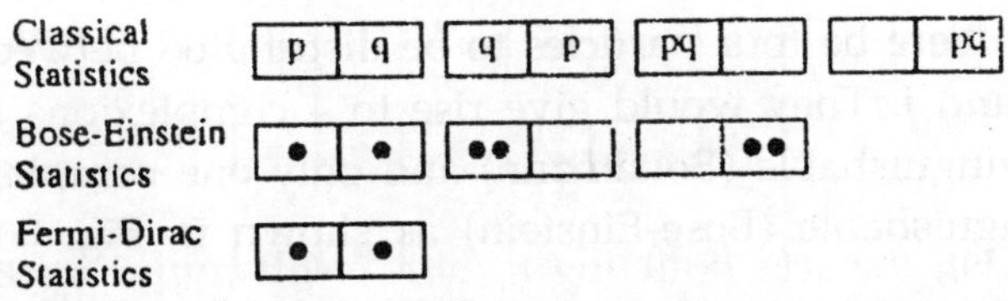

Illustration of essential difference between three statistics.

In Bose-Einstein statistics the particles lose their individuality, hence they are indistinguishable. So we must concentrate our attention on the cell rather than the individual particles. Each cell may have any number of particles, so that in this case we have only three possible arrangements.

In the Fermi-Dirac statistics the particles are indistinguishable and again we have to concentrate our attention on the cells rather than on individual particles. But in this case the particles obey Pauli's exclusion principle according to which it is not possible to have more than one representative point in any one cell. Hence, obviously, each cell will contain one particle. Therefore in this case only one arrangement is possible.

When there are a large number of cells and comparatively few particles to be arranged in them, there is no great difference between the three statistics, because the probability of any cell containing more than one particle is very small. But in a case when the number of particles is comparable with the number of cells, the three statistics lead to divergent results.

Bose-Einstein Statistics

In classical statistics the particles are distinguishable from one another. If two particles interchange their positions (or energy states) a new complexion would occur. But in Bose-Einstein statistics, we start with identical particles which are indistinguishable. Thus the interchange of two particles between two energy states will not give rise to a new complexion or macrostate.

Let there be four particles to be distributed between two cells *i* and *j*. They would give rise to 4 complexions if they are distinguishable (Boltzmann) and only one complexion if indistinguishable (Bose-Einstein) as shown in Fig.

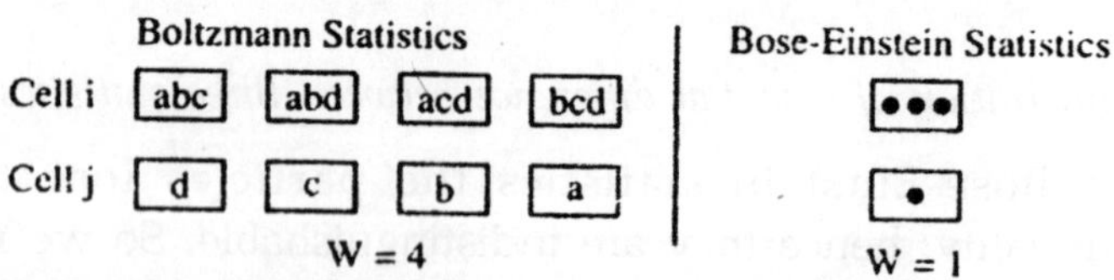

It may be assumed that the available energy states under usual conditions are much larger than the number of molecules.

Let us consider a small energy range so that the molecules may be supposed to have the same energy. The number of energy states even within this range would be quite large. The molecules, due to their indistinguishability, may be distributed in these states without any restriction.

As before, let us consider that there are 4 particles and 2 cells. Let each cell be divided into 4 compartments. Fig. represents the possible distribution without making use of identities of particles. We see that there are different ways of arranging the three phase points in cell *j*. We can, therefore, assign the thermodynamic probability to each cell, equal to the number of possible ways of arranging the particles within that cell. If W_i and W_f represent the probabilities for the respective cells, then in this example W_f = 20, W_i = 4.

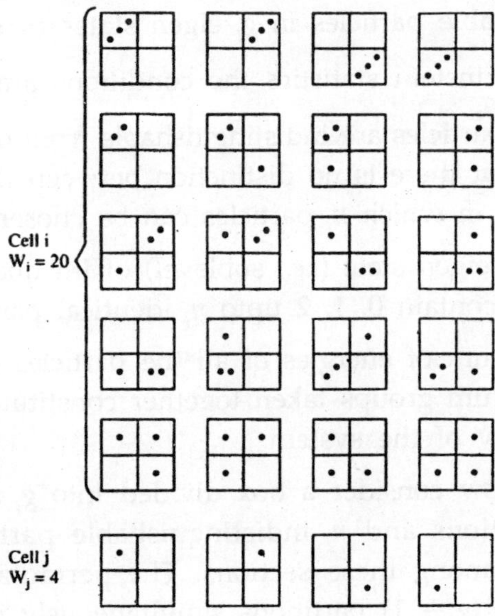

For any one of the arrangements in cell *i*, we can have any one of those in cell *j*, so that the total possible number of arrangements which we call the thermodynamic probability of the macrostate, is

$$W = W_i W_j = 20 \times 4' = 80.$$

This contrasts $W = 4$ on the basis of Maxwell Boltzmann statistics. In general, when there are any number of cells,

$$W = \Pi W_i$$

where the product Π extends over all cells phase space.

In order to derive Bose-Einstein distribution law consider a system of n indistinguishable particles. Let these particles be divided into quantum groups or levels such that there are $n_1, n_2, n_3, \ldots n_i \ldots\ldots$ number of particles in groups whose approximate constant energies are $\varepsilon_1, \varepsilon_2, \varepsilon_3, \ldots\ldots \varepsilon_i \ldots$ respectively. Let g_i be the number of eigen states (*i.e.*, degeneracy or statistical weight) of ith level.

Now we have to ascertain the distribution of n_i indistinguishable particles in g_i eigen states or sublevels.

In Bose-Einstein statistics the conditions are

1. The particles are indistinguishable from one another, so that there is no distinction between the different ways in which n_i particles can be chosen.
2. Each eigen state (*i.e.*, sublevel) of ith quantum state may contain 0, 1, 2 upto n_i identical particles.
3. The sum of energies of all the particles in different quantum groups taken together constitutes the total energy of the system.

Let us now consider a box divided into g_i sections by $(g_i - 1)$ partitions and n_i indistinguishable particles to be distributed among these sections. The permutations of n_i particles and $(g_i - 1)$ partitions simultaneously is given by $(n_i + g_i - 1)!$. But this includes also the permutations of n_i particles among themselves and also $g_i - 1$ partitions among themselves, as both these groups are internally indistinguishable. Hence the actual number of ways in which

n_i particles are to be distributed in g_i sublevels of ith quantum state is

$$\frac{(n_i + g_i - 1)!}{n_i!(g_i - 1)!}.$$

Similar expression can be found for various other quantum states. Therefore the thermodynamic probability will be given by

$$W = \frac{(n_1 + g_1 - 1)!}{n_1!(g_1 - 1)!} \cdot \frac{(n_2 + g_2 - 1)!}{n_2!(g_2 - 1)!} \cdots\cdots \frac{(n_i + g_i - 1)!}{n_i!(g_i - 1)!} \cdots\cdots$$

$$= \Pi \frac{(n_i + g_i - 1)!}{n_i!(g_i - 1)!}. \qquad \text{... (1)}$$

As n_i and g_i are large numbers, we may neglect 1 in above expression, so that

$$W = \prod_i \frac{(n_i + g_i)!}{n_i!(g_i)!}. \qquad \text{... (2)}$$

As n_i and g_i are large numbers we can use Stirling approximation

$$\log n! = n \log n - n!,$$

to obtain $\log W = \sum_i [(n_i + g_i) \log (n_i + g_i) - n_i \log n_i - g_i \log g_j]$

Remembering that g_i is not subject to variation and n_i varies continuously, the differentiation to above equation gives

$$\delta (\log W) = \sum_i [\{\log (n_i + g_i) - \log n_i\}3.\delta n_i]$$

$$= -\sum_i \left\{ \log \frac{n_i}{n_i + g_i} \right\} \delta n_i. \qquad \text{... (3)}$$

For most probable distribution $W = W_{max}$; $\delta (\log W_{max}) = 0$.

Thus the condition of most probable distribution gives

$$\sum_i \left\{ \log \frac{n_i}{n_i + g_i} \right\} \delta n_i = 0. \quad \text{... (4)}$$

The two subsidiary conditions are

(i) Total number of particles in the system is constant

$$n = \sum_i n_i = \text{constant}$$

or

$$\delta E = \sum_i \delta n_i = 0. \quad \text{... (5)}$$

(ii) Total energy of the system is constant

$$\text{E} = \sum_i n_i \ \varepsilon_i = \text{constant}$$

i.e.,

$$\delta \text{E} = \Sigma \varepsilon_\text{i} \ \delta n_i = 0. \quad \text{... (6)}$$

Now let us apply the Lagrangian method of undetermined multipliers. For this, multiplying equation (5) by α, eqn. (6) by β and adding the resulting expressions to eqn. (4) to get

$$\sum_i \left[\log\left(\frac{n_i}{n_i + g_i} \right) + \alpha + \beta\varepsilon_i \right] \delta n_i = 0. \quad \text{... (7)}$$

As the variations Sni are independent of each other, we get

$$\log\left(\frac{n_i}{n_i + g_i} \right) + \alpha + \beta\varepsilon_i = 0$$

or

$$\frac{n_i}{n_i + g_i} = e^{-(\alpha + \beta\varepsilon_i)}$$

i.e.,

$$1 + \frac{g_i}{n_i} = e^{\alpha + \beta\varepsilon_i}$$

i.e.,

$$\frac{g_i}{n_i} = e^{\alpha + \beta\varepsilon_i}$$

i.e.,

$$n_i = \frac{g_i}{(Ce^{\alpha + \beta\varepsilon_i} - 1)} \quad \text{... (8)}$$

This equation represents the most probable distribution of the particles among various energy levels for a system obeying Bose-Einstein distribution law and is, therefore, known as Bose-Einstein distribution law.

Fermi Dirac Statistics

This statistics is obeyed by indistinguishable particles of half-integral spin that have antisymmetric wave functions and obey Pauli's exclusion principle.

Consider a system having n indistinguishable particles. Let these particles be divided into quantum groups or levels such that there are $n_1, n_2, \ldots n_i \ldots$ number of particles in groups whose approximate constant energies are $\varepsilon_1, \varepsilon_2, \ldots \varepsilon_i \ldots$ respectively. Let g_i denote the *degeneracy or statistical weight* (*i.e.*, number of sub–levels) of ith level.

Now the problem is to find the Fermi-Dirac distribution law, *i.e.*, to ascertain the distribution of n_i Fermi particles in g_i eigen states or sub-levels.

In Fermi-Dirac statistics the conditions are:

1. The particles are indistinguishable from each other so that there is no distinction between the different ways in which n_i particles are chosen.
2. The particles obey Pauli's exclusion principle according to which each sub-level or cell may contain 0 to 1 particle. Then, obviously, g_i must be greater than or equal to n_i.
3. The sum of energies of all the particles in the different quantum groups taken together constitutes the total energy of the system.

Now the distribution of n_i particles in g_i states can be done in the following way.

Since due to Pauli's principle no cell can occupy more than one particle, therefore among g_i cells only n_i cells are occupied by one particle each and the remaining $(g_i - n_i)$ cells are empty. The possible number of such a distribution is given by g_i! corresponding to the permutations of the g_i cells. But as the particles are indistinguishable, therefore all the occupied g_i cells are similar to each other. Hence n_i ! permutations of occupied cells will not give rise to indistinguishable arrangements. Further, $(g_i - n_i)$! permutations of empty cells among, themselves also give rise to indistinguishable arrangements. Hence the number of distinguishable arrangements of n_i particles in g_i cells is

$$\frac{g_i!}{n!(g_i n_i)!}. \qquad \text{... (1)}$$

Similar expressions can also be found for various other quantum states. Considering all the available quantum groups, with n_i particles in first group of energy ε_1, n_2 particles in second group of energy ε_2 and so on and keeping in mind that due to indistinguishability of particles, the division of total number of n particles into groups of n_1, n_2 ... n_i particles can be done in only one way; the total number of eigen states fox the whole system is given by

$$G = \prod_i \frac{g_i!}{n_i!(g_i - n_i)!}. \qquad \text{... (2)}$$

In accordance with the postulate of equal a priori probability of states, the probability W of the system for occurring with the specified distribution is proportional to the total number of eigen states, *i.e.*,

$$W = \prod_i \frac{g_i!}{n_i!(g_i - n_i)!}. \times \text{constant}. \qquad \text{... (3)}$$

The Fermi-Dirac distribution law can now be obtained by determining the most probable distribution.

Taking log of eqn. (3), we get

$$\log W = \log \prod_i \left[\frac{g_i!}{n_i!(g_i - n_i)!}. \times \text{constant} \right].$$

$$= \Sigma\left[\log g_i! - \log n_i! - \log(g_i - n_i)!\right] + \text{constant}. \quad \text{... (4)}$$

As n_i and g_i are large numbers, therefore using Stirling approximation, eqn. (4) reduces to

$$\log W = \Sigma\left[g_i \log g_i - g_i - n_i \log n_i + n_i - (g_i - n_i)\log(g_i - n_i) + (g_i - n_i)\right] + \text{constant}$$

$$= \sum_i\left[(n_i - g_i)\log(g_i - n_i) + g_i \log g_i - n_i \log n_i + \text{constant}\right] \quad \text{...(5)}$$

Remembering that g_i is not subject to variation and n_i varies continuously, the differentiation of above equation gives

$$\delta \log W = \sum_i\left\{\log(g_i - n_i)\log n_i\right\}\delta n_i$$

$$= -\sum_i \log\left\{\frac{n_i}{g_i - n_i}\right\}\delta n_i \quad \text{... (6)}$$

For most probable distribution $W = W_{max}$; so, $\delta(\log W_{max}) = 0$.

or $$\sum_i\left\{\log\frac{n_i}{g_i - n_i}\right\}\delta n_i = 0. \quad \text{... (7)}$$

The two subsidiary conditions are

(i) The total number of particles of the system is constant

i.e., $$n = \sum_i n_i = \text{constant}.$$

i.e., $$\delta n = \Sigma\, \delta n_i = 0. \quad \text{... (8)}$$

(ii) Total energy of the system is constant, *i.e.,*

$$E = \sum_i n_i \varepsilon_i = \text{constant}$$

i.e. $$\delta E = \sum_i \varepsilon_i \delta n_i = 0. \quad \text{... (9)}$$

Now to apply the Lagrangian method of undermined multipliers, we multiply eqn. (8) by α and (9) by β and add the resulting expressions to eqn. (7): so that we get

$$\sum_i \left[\log\left(\frac{n_i}{g_i - n_i}\right) + \alpha + \beta\varepsilon_i \right] \delta n_i = 0. \qquad \text{... (10)}$$

As the variations δn_i are independent of each other, we get

$$\log \frac{n_i}{g_i - n_i} + \alpha + \beta\varepsilon_i = 0.$$

or
$$\frac{n_i}{g_i - n_i} = e^{-(\alpha + \beta\varepsilon_i)}$$

ie.
$$\frac{g_i - n_i}{n_i} = e^{\alpha + \beta\varepsilon}$$

ie.
$$\frac{g_i}{n_i} - 1 = e^{\alpha + \beta\varepsilon_i}$$

ie.
$$\frac{g_i}{n_i} = e^{\alpha + \beta\varepsilon_i} + 1$$

ie.
$$n_i = \frac{g_i}{e^{\alpha + \beta\varepsilon_i} + 1}. \qquad \text{... (11)}$$

This equation represents the most probable distribution of the particles among various energy levels for a system obeying Fermi-Dirac statistics and is, therefore, known as Fermi-Dirac distribution law.

Results and Comparision of Three Statistics

The expressions for the most probable distributions in the three statistics are

(i) *Maxwell–Boltzmann Statistics:*

$$n_i = \frac{g_i}{e^{\alpha + \beta\varepsilon_i}}$$

ie.
$$\frac{g_i}{n_i} = e^{\alpha + \beta\varepsilon_i} \qquad \text{... (1)}$$

(ii) *Bose-Einstein Statistics:*

$$n_i = \frac{g_i}{e^{\alpha + \beta\varepsilon_i} - 1}$$

ie. $$\frac{g_i}{n_i}+1 = e^{\alpha+\beta\varepsilon_i} \qquad \text{... (2)}$$

(iii) *Fermi-Dirac Statistics:*

$$n_i = \frac{g_i}{e^{\alpha+\beta\varepsilon_i}+1}$$

ie. $$\frac{g_i}{n_i}-1 = e^{\alpha+\beta\varepsilon_i} \qquad \text{... (3)}$$

It is to be noted that if $\frac{g_i}{n_i}$ is very large in comparison to unity, we may write

$$\frac{g_i}{n_i} \approx \frac{g_i}{n_i}+1 \approx \frac{g_i}{n_i}-1,$$

i.e. for large value of $\frac{g_i}{n_i}$ Bose-Einstein and Fermi-Dirac distributions approach the Maxwell-Boltzmann distribution. This is the case for normal existence of gases when the temperature is not too low and pressure is not too high. Hence classical statistics is capable of describing their behaviour quite adequately. Only under certain extreme conditions the classical statistics fails. The examples of these cases are radiation, liquid Helium II and electron gas inside metals. In all such cases it is essential to use the respective distribution law.

PROBLEMS

1. *Which of the statistics will you use for the systems having: (i) electrons, (ii) photons, (iii) mesons, (iv) oxygen molecules, (v) phonons, (vi) holes, (vii) neutrons, (viii) protons, (ix) He* ^{4} ***atom at low temperature, (x) α–particle, (xi) positron.***

Solution:

(i) Electron, Proton, Neutron, Positron and holes are spin half particles. They are, therefore, fermions and hence obey Fermi-Dirac statistics.

(ii) Photons, phonons, mesons, α-particle and Helium atom at low temperature are particles of integral spin. Therefore, they are Bosons and obey Bose–Einstein statistics.

(iii) Oxygen molecules are classical particles, they obey Maxwell-Boltzmann statistics.

2. ***Classify the following particles according to B-E and F-D statistics:***

Proton, neutron, electron, photon, a–particle, Hydrogen atom, Hydrogen molecule, positron, lithium ion $\left({}^{6}_{3}Li^{++}\right)$

Solution: The atoms or molecules having an even number of fundamental particles have integral spin value and are called Bosons. Similarly, the atoms or molecules having odd number of fundamental particles have half integral spin value and are called Fermions. Accordingly, the classification of particles is given below:

Proton, Neutron, electron and positron are fundamental particles having half spin value. Hence they all are Fermions.

Photon has spin 1 and hence it is Boson

α-particle $({}_{2}He^{4} = 2p + 2n + 2e \rightarrow$ Even $\rightarrow$ Boson

Hydrogen–atom $= 1p + 1n + 1e \rightarrow$ odd $\rightarrow$ Fermion

Hydrogen–molecule $= 2p + 2n + 2e \rightarrow$ Even $\rightarrow$ Boson

Positron $= e^{+} \rightarrow$ odd $\rightarrow$ Fermion

Lithium ion $\left({}^{7}_{3}Li^{++}\right) = 3p + 3n + 2e \rightarrow$ Even $\rightarrow$ Boson

3. ***Three particles are to be distributed in four energy levels a, b, c and d. Write down all the possible ways of this distribution when particles are:***

(i) Fermions (ii) Bosons (iii) Classical Particles.

Solution: Fermions: These particles are indistinguishable having half integral spin. They obey F-D statistics according to which only one particle can occupy one cell. Accordingly, different ways of distribution of $n_i = 3$ particles in $g_i = 4$ cells

$$= \frac{g_i}{n_i!(g_i - n_i)!} = \frac{g_i}{3!(4-3)!} = 4$$

These four ways are shown below:

	a	b	c	d
1.	•	•	•	
2.	•	•		•
3.	•		•	•
4.		•	•	•

Bosons: These are identical indistinguishable particles which do not obey Pauli's exclusion principle. They obey B-E statistics according to which any number of particles can be in any cell (quantum state) and that all quantum states are equally probable. Thus, number of ways in which $n_i = 3$ indistinguishable particles can be distributed among $g_i = 4$ cells

$$= \frac{g_i(n_i + g_i - 1)!}{n_i!g_i!}$$

$$= \frac{4(4+3-1)!}{3!4!}$$

$$= \frac{4 \times (6)!}{3!4!} = \frac{4 \times 5 \times 6}{2 \times 3} = 20$$

These 20 ways are shown below:

	a	b	c	d
1.	•••			
2.		•••		

3.			•••	
4.				•••
5.	••	•		
6.	••		•	
7.	••			•
8.	•	••		
9.		••	•	
10.		••		•
11.	•		••	
12.		•	••	
13.			••	•
14.	•			••
15.		•		••
16.			•	••
17.	•	•	•	
18.	•	•		•
19.	•		•	•
20.		•	•	•

Classical Particles: These are identical particles of any spin and can be distinguished from one another. They obey Maxwell-Boltzmann statistics according to which any number of particles can be in any cell. If we denote the particles by p, q, r then different ways of their distribution in four energy levels

$$= g_i^{n_i} = (4)^3 = 64$$

These 64 ways are shown below:

	a	b	c	d
1.	*pqr*			
2.		*pqr*		

3.			*pqr*	
4.				*pqr*
5.	*pq*	*r*		
6.	*pq*		*r*	
7.	*pq*			*r*
8.	*qr*	*p*		
9.	*qr*		*p*	
10.	*qr*			*p*
11.	*pr*	*q*		
12.	*pr*		*q*	
13.	*pr*			*q*
14.	*r*	*pq*		
15.		*pq*	*r*	
16.		*pq*		*r*
17.	*p*	*qr*		
18.		*qr*	*p*	
19.		*qr*		*p*
20.	*q*	*pr*		
21.		*pr*	*q*	
22.		*pr*		*q*
23.	*r*		*pq*	
24.		*r*	*pq*	
25.			*pq*	*r*
26.	*p*		*qr*	
27.		*p*	*qr*	
28.			*qr*	*p*
29.	*q*		*pr*	
30.		*q*	*pr*	

31.			*pr*	*q*
32.	*r*			*pq*
33.		*r*		*pq*
34.			*r*	*pq*
35.	*p*			*qr*
36.		*p*		*qr*
37.			*p*	*qr*
38.	*q*			*pr*
39.		*q*		*pr*
40.			*q*	*pr*
41.	*p*	*q*	*r*	
42.	*p*	*q*		*r*
43.	*p*	*r*	*q*	
44.	*p*	*r*		*q*
45.	*p*		*q*	*r*
46.	*p*		*r*	*q*
47.	*q*	*p*	*r*	
48.	*q*	*p*		*r*
49.	*q*	*r*	*p*	
50.	*q*	*r*		*p*
51.	*q*	*r*		*p*
52.	*q*		*p*	*r*
53.	*r*	*p*	*q*	
54.	*r*	*p*		*q*
55.	*r*	*q*	*p*	
56.	*r*	*q*		*p*
57.	*r*		*p*	*q*
58.	*r*		*q*	*p*
59.		*p*	*q*	*r*

60.		*p*	*r*	*q*
61.		*q*	*p*	*r*
62.		*q*	*r*	*p*
63.		*r*	*p*	*q*
64.		*r*	*q*	*p*

Application of Bose Einstein Statistics to Planck's Oscillator—Planck's Radiation Law: This example is an application of Bose-Einstein statistics to photons. Radiation from a black-body at absolute temperature T, and in thermal equilibrium is supposed to consist of light quanta or photons of energy content hv, moving in all possible directions with speed of light c and therefore, possesses the momentum $\frac{hv}{c}$.

Some of the important properties of the photons are:

(i) Photons are particles of zero rest masses.

(ii) Photons are indistinguishable from one another and their number in a system is not necessarily constant; because in every emission process in an atom a new light quantum is formed and if a photon of frequency v is absorbed by the wall of the enclosure, it can be replaced by the emission of several photons of frequencies v_1, v_2 ..., provided the total energy of the system is constant, *i.e.*

$$hv = hv_1 + hv_2 + \ldots\ldots$$

Thus in this case, we have

$$\sum_i \delta n_i \neq 0. \qquad \ldots (1)$$

Hence we must drop the auxiliary condition $\sum_i \delta n_i = 0$, and therefore, the undetermined multiplier α is equal to zero.

(iii) The photons are Bose particles with spin 1, having two modes of propagation.

According to quantum idea each allowed eigen state of a quantum mechanical system has a volume h^3 in the phase space. The volume of each allowed eigen state, (*i.e.* each elementary cell in the phase space) can be written as

$$d\tau = dx\, dy\, dz\, dp_x\, dp_y\, dp_z = h^3. \qquad \text{... (2)}$$

Let us now split the phase space into the position space and the momentum space. If we go on increasing the range of position coordinates, till they embrace the whole volume V of the enclosure, then the particles may be found anywhere in the enclosure. So, according to Heisenberg's uncertainty relation an element of volume in the momentum space can be written as

$$\sigma_P = \frac{h^3}{V}. \qquad \text{... (3)}$$

As σ_p denotes the size of an elementary cell in the momentum space, only a single value of momentum can be recognised within a cell. Therefore, this represents an eigen state. If we take an arbitrary portion of the momentum space, only a finite value of momentum, (*i.e.* eigen state) can occur in it. At any instant, all photons having their momenta between p and $p + dp$ will lie within a spherical shell described in momentum space with radii p and $p + dp$. The volume of this shell is $4\pi p^2\, dp$. Therefore, the total number of eigen states between momenta p and $p + dp$ *is* given by

$$g(p)dp = \frac{4\pi p^2 dp}{\frac{h^3}{V}} = \frac{4\pi p^2 V}{h^3} dp. \qquad \text{...(4)}$$

For a photon,

$$p = \frac{hv}{c} \text{ or } dp = \frac{h\, dv}{c}.$$

Substituting these values in equation (4), the total number of eigen states between frequencies v and $v + dv$ *is* given by

$$g(v)\,dv = 4\pi V \cdot \frac{v^2}{c^3} dv. \qquad \text{... (5)}$$

Taking into account the doubling of the states due to polarisation of the photons, (*i.e.* two modes of propagation for each photon) the total number of eigen states available for the photons in the frequency range v and $v + dv$ *is* given by

$$g(v)\,dv = 8\pi V \cdot \frac{v^2}{c^3} dv. \qquad \text{... (6)}$$

In this case $\alpha = 0$, $\beta = \frac{1}{kT}$ and $\varepsilon = h\nu$ (the energy of a photon), therefore, using (6) above equation may be written as

$$dn = 8\pi V \cdot \frac{v^2}{c^3} \cdot \frac{dv}{e^{hv/kT} - 1} \qquad \text{...(7)}$$

i.e.
$$\frac{dn}{V} = \frac{8\pi v^2}{c^3} \cdot \frac{dv}{e^{hv/kT} - 1} \qquad \text{...(8)}$$

This equation represents the number of photons per unit volume lying in the frequency range v and $v + dv$.

The energy density of radiation of frequencies between v and $v + dv$ can now be found by multiplying equation (8) by the energy of the photon hv. Therefore, if $u_v\, dv$ represents the energy density of radiation within the specified frequency range, then we get the energy distribution law

$$u_v dv = \left(\frac{dn}{V} hv\right) = \left(\frac{8\pi v^2 dv}{c^3}\right) \cdot \left(\frac{hv}{e^{hv/kT} - 1}\right)$$

i.e.
$$u_v dv = \frac{8hv^3}{c^3} \cdot \frac{dv}{e^{hv/kT} - 1} \qquad \text{... (9)}$$

This is well known Planck's law of radiation in terms of frequency. Equation (9) in terms of wavelength λ becomes

$$\left(\text{Since } v = \frac{c}{\lambda} \text{ and } dv = -\frac{cd\lambda}{\lambda^2}\right)$$

$$u_\lambda d\lambda = \frac{8\pi h}{\lambda^3} \cdot \frac{\left(-cd\lambda / \lambda^2\right)}{e^{hc/\lambda kT} - 1}$$

i.e.

$$\left|u_\lambda d\lambda\right| = \frac{8\pi hc}{\lambda^5} \cdot \frac{d\lambda}{e^{hc/\lambda kT} - 1}. \qquad \text{... (10)}$$

This is well known Planck's law of radiation in terms of wave length.

Hence the Bose-Einstein statistics while confirming the validity of Planck's radiation law, has the merit of using only photons and no other hypothetical resonators in the deduction of the law governing the back-body radiation.

Bose-Einstein Gas

It may be pointed out that Bose-Einstein and Fermi-Dirac distributions need not be used for ordinary atomic or molecular gases; since these gases are in practice, hence they are, always described with sufficient accuracy by the Boltzmann distribution. An assembly of bosons (*i.e.* indistinguishable elementary particles of zero or integral spin) is termed as Bose-Einstein gas.

Consider a perfect Bose-Einstein gas of n bosons. Let these particles be distributed among quantum groups or states such that there are n_1, n_2, ..., n_i, ... number of particles in quantum states whole approximate constant energies are ε_1, ε_2, ..., ε_i respectively.

As the gas is assumed to be perfect, the interaction between its particles is negligibly so that the energy may be regarded as entirely translational in character. Therefore, the result obtained in this case will be particularly applicable to the mono–atomic gas.

If g_i is the degeneracy or statistical weight of ith quantum states, then according to Bose-Einstein distribution law of most probable distribution is given by

$$n_i = \frac{g_i}{e^{\alpha+\beta\varepsilon_i} - 1} \qquad \text{... (1a)}$$

where $$\alpha = \frac{\mu}{kT} \text{ and } \beta = \frac{1}{kT}.$$

Equation (la) may be written as

$$n_i = \frac{g_i}{\frac{1}{A}e^{\beta\varepsilon_i} - 1} \qquad \text{... (1b)}$$

where $A = e^{-\alpha}$. ... (2)

As the number of particles in a state cannot be negative, we must always have $n_i \geq 0$. This requires

$$e^{\alpha+\beta\varepsilon_i} = \frac{1}{A}e^{\beta\varepsilon_i} \geq 1. \qquad \text{...(3)}$$

If the ground state (or lowest energy state) is taken to be at zero energy, then

$$e^{\alpha} = e^{-\mu/\kappa T} = \frac{1}{A} \geq 1. \qquad \text{... (4)}$$

i.e., $$\alpha \geq 0;\ \mu \geq 0,\ 0 \leq A \leq 1. \qquad \text{... (5)}$$

The constant α can be determined by the condition

$$n = \sum_i n_i = \sum_i \frac{g_i}{e^{\beta\varepsilon_i} - 1}.$$

$$= \sum_i \frac{g_i}{\frac{1}{A}.e^{\beta\varepsilon_i} - 1}. \qquad \text{... (6)}$$

It is known that for the particles in a box of normal size, the translational levels are spaced closely enough so that we can integrate over phase space instead of summing over particle states.

The number of particle states $g(p)\,dp$ lying between momentum p and $p + dp$ is given by

$$g(p)dp = \frac{4\pi p^2 dp}{h^3/V} \qquad \text{... (7)}$$

where g_s is the spin degeneracy factor caused by the particle spin s.

For simplicity let us take $s = 0$ *i.e.* $g_s = 1$, then eqn. (7) becomes

$$g(p)dp = g_s \frac{4\pi V}{h^3} p^2 dp \qquad \text{... (8)}$$

According to eqn. (1) for most probable distribution the number of particles in the momentum range p and $p + dp$ is given by

$$dn(p) = \frac{g(p)dp}{e^{\alpha+\beta\varepsilon_i} - 1} \qquad \text{... (9)}$$

Now, remembering that for perfect gas (non-interacting) particles

$$\varepsilon = \frac{p^2}{2m} \qquad \text{... (10)}$$

and using eqn. (8), we get

$$dn(p) = \frac{4\pi V}{h^3} \cdot \frac{p^2 dp}{e^{\alpha + p^2/2mkT} - 1} \left\{ \text{since } \beta = \frac{1}{kT} \right\} \qquad \text{... (11)}$$

Using eqns. (10) and (11) the number of particles lying between energy range ε and $\varepsilon + de$ may be written as

$$dn(\varepsilon) = \frac{4\pi V}{h^3} \cdot \frac{(2m\varepsilon)\left\{\frac{m}{2\varepsilon}\right\}^{1/2} d\varepsilon}{e^{\alpha+\varepsilon/kT} - 1}$$

$$= \frac{4\pi m V}{h^3} \sqrt{(2m)} \frac{\varepsilon^{1/2} d\varepsilon}{e^{\alpha+\varepsilon/kT} - 1} \qquad \text{... (12)}$$

Now let us substitute

$$x = \frac{\varepsilon}{kT} \text{ or } dx = \frac{d\varepsilon}{kT} \qquad \text{... (13)}$$

So that eqn. *(12)* may be expressed as

$$dn(\varepsilon) = \frac{4\pi m V}{h^3} (2m)^{1/2} \cdot \frac{(kTx)^{1/2}(kTdx)}{e^{\alpha+x} - 1}$$

$$= \frac{V}{h^3} (2\pi mkT)^{3/2} \frac{2}{\sqrt{\pi}} \cdot \frac{x^{1/2} dx}{e^{\alpha+x} - 1} \qquad \text{... (14)}$$

Now, keeping in mind the closely spaced levels in this case, the total number of particles is given by

$$n = \int dn(\varepsilon)$$

$$= \frac{V}{h^3}(2\pi\, m\, kT)^{3/2} \frac{2}{\sqrt{\pi}} \int_0^\infty \frac{x^{1/2}dx}{e^{\alpha+x}-1}$$

$$= \frac{V}{h^3}(2\pi\, m\, kT)^{3/2} f_1(\alpha) \qquad \text{... (15)}$$

where
$$f_1(\alpha) = \frac{2}{\sqrt{\pi}} \int_0^\infty \frac{x^{1/2}dx}{e^{\alpha+x}-1} \qquad \text{... (16)}$$

The total energy of *the system is given* by

$$E = \int \varepsilon dn(\varepsilon) = \int kTx \,.\, dn(\varepsilon) \text{ using (13)}$$

$$= \frac{V}{h^3}(2\pi mkT)^{3/2} kT . \frac{2}{\sqrt{\pi}} \int_0^\infty \frac{x^{3/2}dx}{e^{\alpha+x}-1} \text{using (14)}$$

$$= \frac{3}{2} . \frac{V}{h^3}(2\pi mkT)^{3/2} kT f_2(\alpha) \qquad \text{... (17)}$$

where
$$f_2(\alpha) = \frac{4}{3\sqrt{\pi}} \int_0^\infty \frac{x^{1/2}dx}{e^{\alpha+x}-1}. \qquad \text{... (18)}$$

For A < 1 the integral functions f_1 (α) and f_2 (α) may be evaluated by expanding the series as follows:

$$f_1(\alpha) = \frac{2}{\sqrt{\pi}} \int_0^\infty \frac{x^{1/2}dx}{e^{\alpha+x}-1} = \frac{2}{\sqrt{\pi}} \int_0^\infty x^{1/2} \left(\frac{e^x}{A}-1\right)^{-1} dx$$

$$= \frac{2}{\sqrt{\pi}} \int_0^\infty x^{1/2} A e^{-x} \left(1 - Ae^{-x}\right)^{-1} dx$$

$$= \frac{2}{\sqrt{\pi}} \int_0^\infty x^{1/2} A e^{-x} \left(1 + Ae^{-x} + A^2 e^{-2x} + ...\right) dx$$

$$= \frac{2}{\sqrt{\pi}} \left[A \int_0^\infty x^{1/2} e^{-x} dx + A^2 \int_0^\infty x^{1/2} e^{-2x} dx + ... \right]$$

$$= \frac{2}{\sqrt{\pi}} \left[\frac{\sqrt{\pi}}{2} \left(A + \frac{A^2}{2^{3/2}} + \frac{A^3}{3^{3/2}} + ... \right) \right]$$

$$= A + \frac{A^2}{2^{3/2}} + \frac{A^3}{3^{3/2}} + ... = \sum_{r=1}^{\infty} \frac{A^r}{r^{3/2}} \qquad \text{... (19)}$$

Similarly we can write

$$f_2(\alpha) = A + \frac{A^2}{2^{5/2}} + \frac{A^3}{3^{5/2}} + \ldots\ldots = \sum_{r=1}^{\infty} \frac{A^r}{r^{3/2}} \quad \ldots (20)$$

Therefore the expressions for total number of particles and total energy of perfect Bose-Einstein gas become

$$n = \frac{V}{h^3}(2\pi mkT)^{3/2} \sum_{r=1}^{\infty} \frac{A^r}{r^{3/2}}$$

$$= \frac{V}{h^3}(2\pi mkT)^{3/2}\left(A + \frac{A^2}{2^{3/2}} + \frac{A^3}{3^{3/2}} + \ldots\right) \quad \ldots (21)$$

$$E = \frac{3V}{2h^3}(2\pi mkT)^{3/2} .kT \sum_{r=1}^{\infty} \frac{A^r}{r^{5/2}}$$

$$= \frac{3V}{2h^3}(2\pi mkT)^{3/2} .kT\left(A + \frac{A^2}{2^{5/2}} + \frac{A^3}{3^{5/2}} + \ldots\right) \quad \ldots (22)$$

Dividing eqn. (22) by (21), we get

$$\frac{E}{n} = \frac{3}{2}kT\left(A + \frac{A^2}{2^{5/2}} + \frac{A^3}{3^{5/2}} + \ldots\right)\left(A + \frac{A^2}{2^{3/2}} + \frac{A^3}{3^{3/2}} + \ldots\right)^{-1}$$

i.e.

$$E = \frac{3}{2}nkT\left[1 - \frac{A^2}{2^{5/2}} - \frac{A^3}{3^{5/2}} - \ldots\ldots\right] \quad \ldots (23)$$

The value of α or A can be determined by eqn. *(15)* or *(21)*, *i.e.*

$$f_1(\alpha) = \frac{n}{V} \cdot \frac{h^3}{(2\pi mkT)^{3/2}} \quad \ldots (24)$$

Obviously, f_1 (α) is proportional to the particle density $\frac{n}{V}$ and inversely proportional to $T^{3/2}$.

For A << 1, f_1 (α) = A, so eqn. (19) gives

$$A = e^{-\alpha} \approx \frac{n}{V} \cdot \frac{h^3}{(2\pi mkT)^{3/2}} \quad \ldots (25)$$

Obviously, A would be small for high temperatures or low density (or classical limit).

However, the general value of A or α can be obtained by inverting the series for f_1 (α)

$$f_1(\alpha) = A + 2^{-3/2} A_2 + 3^{-3/2} A_3 + \ldots\ldots$$

or $$A = e^{-\alpha} = f_1(\alpha) - 2^{-3/2} f_1^2(\alpha) + (2^{-3/2} - 3^{-3/2}) f_1^3(\alpha) \qquad \ldots (26)$$

This equation represents value of A as a function of f_1 (α).

(a) Maxwell-Boltzmann distribution as α limiting case of Bose-Einstein distribution.

For $A << 1, e^{\alpha+\beta\varepsilon_i}$ i.e. $\frac{1}{A}e^{\beta\varepsilon_i}$ becomes very large in comparison to 1, so that Bose-Einstein distribution

$$n_i = \frac{g_i}{e^{\alpha+\beta\varepsilon_i} - 1}$$ may be written as

$$n_i \approx \frac{g_i}{e^{\alpha+\beta\varepsilon_i}},$$

which is Maxwell–Boltzmann distribution. Thus in the limiting case of A << 1, the Bose–Einstein distribution law reduces to Maxwell-Boltzmann distribution.

When A << 1, higher order terms in A may be neglected and then

$$f_1(\alpha) = f_2(\alpha) = A.$$

So that we may write eqn. (21) and (22) as

$$n = \frac{V}{h^3}(2\pi m\, kT)^{3/2}.A, \qquad \ldots (27)$$

$$E = \frac{3}{2} \cdot \frac{V}{h^3}(2\pi m\, kT)^{3/2}. kT.\ A, \qquad \ldots (28)$$

Dividing eqn. (28) by (27), we get

$$\frac{E}{n}=\frac{3}{2}kT.\ \text{i.e.}\ E=\frac{3}{2}.\,nkT, \quad \text{... (29)}$$

which is well-known result of classical (Maxwell-Boltzmann) Statistics.

(b) Degeneracy and Bose-Einstein Condensation

From eqn. (24), the approximate value of A is

$$A \ = e^{-\alpha} \approx \frac{n}{V}.\frac{h^3}{(2\pi mkT)^{3/2}} \quad \text{... (30)}$$

If the density of particles is increased and/or the temperature is decreased, the value of A increases (or $\alpha = -\mu/kT$ decreases), then the behaviour of the perfect gas departs farther and farther from that of the classical to quantum statistics and not to classical statistics. The gas under this condition is said to be degenerate and the parameter A is called the degeneracy parameter.

As the expression for A contains three variables, *viz.* m, the mass of the particle, $\frac{n}{V}$ the particle density, *i.e.* the number of particles per unit volume and T the absolute temperature of the gas, obviously the criterion of degeneracy will be based on the magnitude of $\frac{n/V}{(mT)^{3/2}}$. Hence the degree of degeneracy will be large when temperature is low, particle density is large and the mass of each boson is small. On account of the above three factors, the gas can become degenerate in three different ways

For low energy values the maximum admissible value of A is 1 [refer eqn. (5)] and consequently α can never be negative. Thus, for low energy values the limiting case of highest degeneracy in Bose–Einstein gas reaches when $A = 1$ and $\alpha = 0$. Then the maximum value of f_1 (α) will be given by

$$[f_1(\alpha)]_{max} = f_1(0) = 1 + \frac{1}{2^{2/3}} + \frac{1}{3^{3/2}} + ... = 2.612 \quad ...(31)$$

So the maximum value of particle density $\frac{n}{V}$ will be given by

$$\left(\frac{n}{V}\right)_{max} = \frac{(2\pi mkT)k^{3/2}}{h^3}(2.612) \quad ... (32)$$

Since eqn. (32) corresponds to the limiting case of the Bose-Einstein degeneration, no solution of eqn. (15) can exist for

$$\frac{n}{V} > \frac{(2\pi mkT)^{3/2}}{h^3}.2.612 \quad ...(33)$$

because this would involve A > 1.

The fact that no value of $\frac{n}{V}$ can be greater than that given by eqn. (32), can be alternatively expressed in terms of the critical temp. T_0 defined as

$$\frac{n}{V} = \frac{(2\pi mkT_0)^{3/2}}{h^3}.2.612 \quad ... (34a)$$

i.e.

$$T_0 = \frac{h^2}{2\pi mk}\cdot\left(\frac{1}{2.612}\cdot\frac{n}{V}\right)^{2/3} \quad ... (34b)$$

Thus the critical temp. T_0 is the lowest temperature for which a solution of eqn. (15) is possible, *i.e.* there is no solution of equation (15) for T < T_0, T_0 is therefore the temperature at which the degeneracy of energy levels starts. If we plot a graph between the energy *E* and the temperature *T* of the gas, a curve of the type shown in fig. is obtained.

Below the critical temperature T_0 the full line shows the relation between *E* and *T* of a degenerate gas, while the dotted line through the origin that of a non-degenerate gas. E_0 is termed as zero point energy which will be understood in the applications of Fermi-Dirac statistics. The critical temperature T_0 is 5K for helium gas.

Now the question arises why there is no solution of eqn. (15) for $T < T_0$. The reason is that while arriving at Bose-einstein distribution, because of the closeness of energy levels, we have assumed the continuous distribution in place of discrete distribution and hence have replaced the summation by integration; while at low temperatures, the number of particles begins to crowd into lower energy levels and a large number of particles may occupy the ground state $\varepsilon_0 = 0$.

This means that at low temperatures we must be careful in replacing the summation into integration.

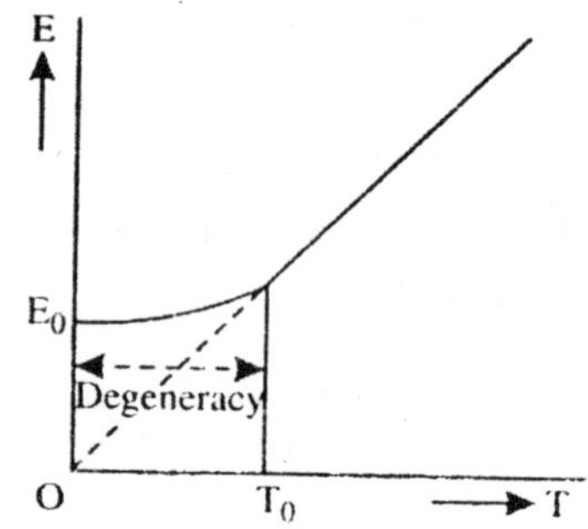

Now from eqn. (12) the number of particles lying between energy range ε and $\varepsilon + d\varepsilon$ is given by

$$dn(\varepsilon) = \frac{g(\varepsilon)d\varepsilon}{e^{\alpha+\varepsilon/kT}} \qquad \text{... (35)}$$

where

$$g(\varepsilon) = \frac{4\pi mV}{h^3}(2m\varepsilon)^{1/2} \qquad \text{... (36)}$$

It may be now noted that for ground state $\varepsilon = \varepsilon_0 = 0$, $g(\varepsilon) = 0$, while actually it should be unity $g(0) = 1$ as there is one state at $\varepsilon = 0$. Therefore the above distribution [eqn. (35)] gives incorrect result for ground state, while this state is very important at low temperature. We further note at $\varepsilon \neq 0$, $g(\varepsilon) \neq 0$ and therefore the above distribution holds good. Consequently the distribution (35) can still be applied for all states except ground state which should be treated separately.

For a single state, we have

$$n_i = \frac{g_i}{\varepsilon^{\alpha+\beta\varepsilon_i} - 1}$$

For ground state $\varepsilon_i = \varepsilon_0 = 0$ and $g_i = 1$. Therefore the number of particles in the ground state is given by

$$n_i \to n_0 = \frac{1}{e^{\alpha} - 1} \quad \text{... (37)}$$

Therefore the total number n of particles for the degenerate case may be expressed as

$$n_i = n_0 + \int dn(\varepsilon) = n_0 + \int_0^{\infty} \frac{4\pi mV}{h^3} \cdot \frac{(2m\varepsilon)^{1/2}}{e^{\alpha+\varepsilon/kT} - 1} d\varepsilon$$

$$= n_0 + n'. \quad \text{... (38)}$$

where $$n' = \frac{4\pi mV}{h^3} \cdot \sqrt{(2m)} \int_0^{\infty} \frac{\varepsilon^{1/2} d\varepsilon}{e^{\alpha+\varepsilon/kT} - 1} \quad \text{... (39)}$$

$$= \frac{V}{h^3} \cdot (2\pi mkT)^{3/2} f_1(\alpha) \text{ using (16)}$$

$$= n\left(\frac{T}{T_0}\right)^{3/2} \frac{f_1(\alpha)}{f_1(0)} \quad \text{...(40)}$$

(using 34a) as $f_1(0) = 2.612$.

As $f_1(\alpha) < f_1(0)$, therefore n' given by eqn. (40) acquires its maximum value when $\alpha = 0$, thus the maximum number of particles (n') occupying states above the ground state is given by

$$n' = n\left(\frac{T}{T_0}\right)^{3/2} \quad (\text{for } T > T_0) \quad \text{... (41)}$$

Therefore the rest of particles, given by

$$n_0 = n - n' = n\left[1 - \left(\frac{T}{T_0}\right)^{3/2}\right], \quad (\text{for } T < T_0) \quad \text{...(42)}$$

must condense into the ground state.

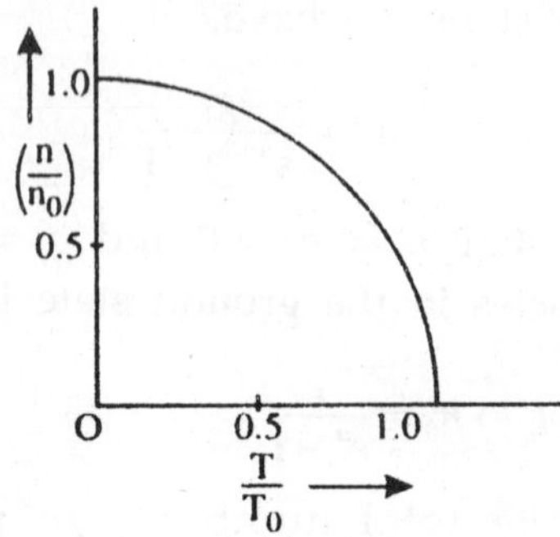

From eqn. (42) it is obvious that when the temperature of a Bose-Einstein gas is lowered below the critical temperature T_0, the number of particles in the ground state rapidly increases. This rapid increase in the population of the ground state below the critical temperature T_0 for a Bose-Einstein gas is called the Bose-Einstein condensation. It is obvious from eqn. (34b) that the critical temperature T_0, at which the Bose-Einstein condensation starts, depends upon the particle density $\frac{n}{V}$ of the gas.

Eqn. (42) is plotted in fig. which represents the fraction of particles condensed in the ground state for $T \le T_0$. At the ground state $\varepsilon = 0$, the particles of a Bose-Einstein degenerate gas condensed in the ground state do not contribute to the energy.

For temperature above T_0, $\alpha \ne 0$, the number of particles is given by eqn. (15) with negligible number of particles in the ground state and the gas is said to be the *classical* or *non-degenerate.*

In the case of Helium in liquid state T_0 can be calculated to have the value 3.12K. Therefore the degeneration and condensation of helium must start at 3.12K. But experimental observations of the lambda-point transition in liquid helium show that the condensation of helium starts at 2.19K which is close to the calculated value. Below this temperature the liquid helium exhibits the remarkable physical properties of a superfluid. It is generally believed that the lambda point transition observed in liquid helium at 2.19K is essentially a Bose-Einstein condensation.

Fermi-dirac Gas

An assembly of fermions (*i.e.*, indistinguishable particles of half integral spin obeying Pauli exclusion principle) is known as *Fermi-Dirac gas.*

Consider a perfect Fermi-Dirac gas of n fermions. Let these particles be distributed among quantum groups of states such that there are $n_1, n_2 \ldots, n_i \ldots$ number of particles in quantum state whose approximate constant energies are $\varepsilon_1, \varepsilon_2, \ldots \varepsilon_i \ldots$ respectively.

If g_i is the degeneracy or statistical weight of ith quantum state, then according to Fermi-Dirac distribution law the most probable distribution is given by

$$n_i = \frac{g_i}{e^{\alpha+\beta\varepsilon_i}+1} \quad \ldots (1a)$$

where $$\alpha = -\frac{\mu}{kT} \text{ and } \beta = \frac{1}{kT}.$$

μ being chemical potential, T absolute temperature and k Boltzmann constant.

Eqn. (la) may be written as

$$n_i = \frac{g_i}{\frac{1}{A}e^{\beta\varepsilon_i}+1} \quad \ldots (1b)$$

where $$A = e^{-\alpha} \quad \ldots (2)$$

As in the denominator of eqn. (1) a positive sign appears before one, α need not be restricted to positive values as in Bose–Einstein case; but may be positive or negative. The number of particle – states lying between momentum p and $p + dp$ *is* given by

$$g(p)dp = g_s.\frac{4\pi Vp^2dp}{h^3/V} = gs.\frac{4\pi p^2dp}{h^3} \quad ...(3)$$

where g_s *is* the spin degeneracy factor caused by the particle spin s and given by

$$g_s = (2s + 1) \quad ...(4)$$

Since $\varepsilon = \frac{p^2}{2m}$ or $dp = \left(\frac{m}{2\varepsilon}\right)^{1/2} d\varepsilon$, the number of states between energy ε and ε + $d\varepsilon$ is given by

$$g(\varepsilon)d\varepsilon = g_s.\frac{4\pi V(2m\varepsilon)\left(\frac{m}{2\varepsilon}\right)^{1/2} d\varepsilon}{h^3} \quad ...(5)$$

For a gas of fermions at temperature T in a volume V, the mean number of fermions in the energy range between ε and ε + $d\varepsilon$ is given by

$$dn(\varepsilon) = g_s.\frac{4\pi mV}{h^3}\sqrt{(2m)}.\frac{\varepsilon^{1/2}d\varepsilon}{\varepsilon^{\alpha+\varepsilon/kT}+1} \quad ...(6)$$

As the gas is assumed to be perfect, the interaction between its particles is negligible, so that the energy may be regarded as entirely translational in character. Further, in a box of normal size the translational levels are spaced closely enough so that we can integrate over phase space instead of summing over particle states.

Therefore the total number of particles in Fermi-Dirac gas is given by

$$n = \Sigma dn(\varepsilon) = \int dn(\varepsilon) = g_s.\frac{4\pi mV}{h^3}\sqrt{(2m)}.\int_0^\infty \frac{\varepsilon^{1/2}d\varepsilon}{e^{\alpha+\varepsilon/kT}+1} \quad ...(7)$$

and the total internal energy is given by

$$E = \int_0^\infty \varepsilon dn(\varepsilon) = g_s.\frac{4\pi mV}{h^3}\sqrt{(2m)}.\int_0^\infty \frac{\varepsilon^{1/2}d\varepsilon}{e^{\alpha+\varepsilon/kT}+1}; \quad ...(8)$$

Now substituting $x = \frac{\varepsilon}{kT}$ or $dx = \frac{d\varepsilon}{kT}$; ...(9)

Equation (7) and (8) take the form

$$n = g_s \cdot \frac{V}{h^3}(2\pi mkT)^{3/2} \cdot \frac{2}{\sqrt{\pi}} \int_0^\infty \frac{x^{1/2}dx}{e^{\alpha+x}+1}$$

$$= g_s \cdot \frac{V}{h^3}(2\pi mkT)^{3/2} f_1(\alpha) \qquad \text{... (10)}$$

where $$f_1(\alpha) = \frac{2}{\sqrt{\pi}} \int_0^\infty \frac{x^{1/2}dx}{\varepsilon^{\alpha+x}+1} = \frac{2}{\sqrt{\pi}} \int_0^\infty \frac{x^{1/2}dx}{\frac{1}{A}e^x+1} \qquad \text{... (11)}$$

and $$E = \frac{3}{2} g_s \cdot \frac{V}{h^3}(2\pi mkT)^{3/2} .kT. \frac{4}{3\sqrt{\pi}} \int_0^\infty \frac{x^{3/2}dx}{e^{\alpha+x}+1}$$

$$E = \frac{3}{2} g_s \cdot \frac{V}{h^3}(2\pi mkT)^{3/2} .kTf_2(\alpha) \qquad \text{... (12)}$$

where $$f_2(\alpha) = \frac{4}{3\sqrt{\pi}} \int_0^\infty \frac{x^{3/2}dx}{e^{\alpha+x}+1} = \frac{4}{3\sqrt{\pi}} \int_0^\infty \frac{x^{3/2}dx}{\frac{1}{A}e^x+1} \qquad \text{... (13)}$$

The integrals f_1 (α) and f_2 (α) must be evaluated for both positive and negative values of α.

It is often convenient to introduce the Fermi–distribution function f (ε) defined by

$$f(\varepsilon) = \frac{n(\varepsilon)}{g(\varepsilon)} = \frac{1}{e^{\alpha+\varepsilon/kT}+1} = \frac{1}{e^{(\varepsilon-\varepsilon_F)/kT}+1} \qquad \text{... (14)}$$

where α in terms of chemical potential μ or Fermi-energy ε_F (T) at temperature T is given by

$$\alpha = -\frac{\mu}{kt} = \frac{\varepsilon_F(T)}{kT} \qquad \text{... (15)}$$

At T = 0K

$$\left.\begin{aligned} f(\varepsilon) &= 1 \text{ for } \varepsilon_F < \varepsilon_F(0) \\ &= 0 \text{ for } \varepsilon_F > \varepsilon_F(0) \end{aligned}\right\} \qquad \text{... (16)}$$

At any temperature T,

$$f\varepsilon = \frac{1}{2} \text{ for } \varepsilon = \varepsilon_F.$$

The value of ε_F is determined by the condition that the total number of particles is constant at a given temperature T.

The expression for the number of fermions with energy between ε and $\varepsilon + d\varepsilon$ is written as

$$dn(\varepsilon) = f(\varepsilon) g(\varepsilon) d\varepsilon = g_s . = \frac{4\pi mV}{h^3} \sqrt{(2m)} . \frac{\varepsilon^{1/2} d\varepsilon}{e^{\varepsilon - \varepsilon_F/kT} + 1} \quad \text{... (17)}$$

Thus at absolute zero all states with $0 < \varepsilon \leq \varepsilon_F(0)$ are completely filled and all states with $\varepsilon > \varepsilon_F$ (0) are empty. The value ε_F (0) of ε_F at T = 0*K is* determined by

$$n = \int_0^{\varepsilon_F(0)} dn(\varepsilon) = \int_0^{\varepsilon_F(0)} f(\varepsilon) g(\varepsilon) d\varepsilon = \int_0^{\varepsilon_F(0)} g(\varepsilon) d\varepsilon$$

$$= \int_0^{\varepsilon_F(0)} g_s . \frac{4\pi mV}{h^3} (2m)^{1/2} \varepsilon^{1/2} d\varepsilon = g_s \frac{4\pi mV}{h^3} (2m)^{1/2} . \frac{2}{3} \left[\varepsilon_F(0)\right]^{3/2}$$

i.e.

$$\varepsilon_F(0) = \frac{h^2}{2m} \left[\frac{3n}{4\pi V g_s}\right]^{3/2} = \mu_F . \quad \text{... (18)}$$

This represents the energy of the height level occupied at T = 0*K*.

For particles of spin $\frac{1}{2}$ (e.g. electrons, protons, etc.), the spin degeneracy factor g_s is 2. This means that each level can be occupied by two particles, one with spin up and the other with spin down $(\uparrow\downarrow)$.

If $\rho = \left\{\frac{mn}{V}\right\}$ kg/m³ *is* the density of the gas, then numerically,

$$\varepsilon_F(0) = \mu_F \begin{cases} 0.625 \times 10^{-17}\ \rho^{2/3} \text{ joule or } 39\ \rho^{2/3} \text{ eV for electrons} \\ 0.227 \times 10^{-20}\ \rho^{2/3} \text{ joule or } 1.42\ \rho^{2/3} \text{ eV for protons.} \end{cases} \cdots (19)$$

For conduction electrons in metals ρ = 0.1 kg/m³. Thus Fermi-Dirac gas possesses appreciable energy at absolute zero, while both Bose-Einstein and classical statistics predict zero value of energy at 0K. This is because n_i can have 0 or 1 only for Fermi-Dirac statistics. As the temperature increases above 4K, the Fermi-Dirac distribution function near ε_F (0) takes the shape of dotted line.

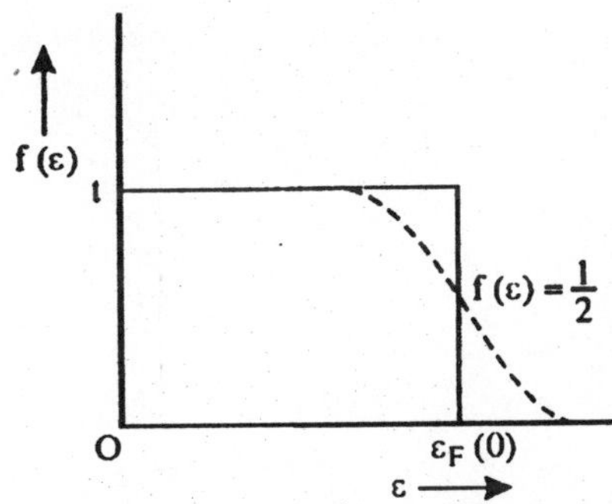

Plot of Fermi-Dirac distribution function at T=0K (solid line) and at low temperature $\frac{kT}{\varepsilon F} \ll 1$ ***(dotted line)***

Let us now define a temperature T_F, called Fermi–temperature, by

$$T_F = \frac{\varepsilon_F(0)}{k} = \frac{\mu_F}{k} = \frac{h^2}{2mk}\left(\frac{3n}{4\pi\, Vg_s}\right)^{2/3} = \frac{h^2}{2m^{5/2}k} \cdot \left(\frac{3\rho}{4\pi g_s}\right)^{2/3} \quad \text{... (20)}$$

When $T \ll T_F$ *or* $kT \ll \varepsilon_F(0)(=kT_F)$, then the distribution is called strongly degenerate. When $T > T_F$ *or* $kT > \varepsilon_F(0)$, then the distribution is called slightly or weakly degenerate. When $T \gg T_F$ *or* $kT \gg \varepsilon_F(0)$, then we get the classical limit and the distribution is non-degenerate.

The value of α is negative in the case of strongly degenerate systems and positive for non-degenerate and slightly degenerate systems. Due to eqn. (16) this means that $\varepsilon_F > 0$ at low temperatures and $\varepsilon_F < 0$ at high temperatures.

zero, while from Bose-Einstein and classical statistics predict zero value of energy at 0K. This is because it can have 0 or 1 only for Fermi-Dirac statistics. As the temperature increases Fermi-Dirac distribution function near $\epsilon_F(0)$ takes the shape of dotted line.

Plot of Fermi-Dirac distribution function at $T=0K$ (solid line) and at low temperature $\frac{kT}{\epsilon_F} \ll 1$ (dotted line)

Let us now define a temperature T_F, called Fermi temperature, by

$$T_F = \frac{\epsilon_F(0)}{k} = \frac{h^2}{2mk}\left(\frac{3n}{4\pi g_s V}\right)^{2/3} = \frac{h^2}{2mk}\left(\frac{3N}{4\pi V}\right)^{2/3} \quad (2)$$

When $T \ll T_F$ or $kT \ll \epsilon_F(0)(=kT_F)$, then the distribution is called strongly degenerate. When $T \simeq T_F$ or $kT \simeq \epsilon_F(0)$, then the distribution is called slightly or weakly degenerate. When $T \gg T_F$ or $kT \gg \epsilon_F(0)$, then we get the classical limit and the distribution is non-degenerate.

The value of α is negative in the case of strongly degenerate systems and positive for non-degenerate and slightly degenerate systems. Due to eqn. (16) this means that $\epsilon_F > 0$ at low temperatures and $\epsilon_F < 0$ at high temperatures.

Elements in Transition

The structures of the products formed in various reactions, keeping in view the 18-electron rule.

$$\eta^4C_4H_6Fe(CO)_3 + HC1 \longrightarrow \eta^4C_4H_6\text{—}HFe(CO)_3Cl$$

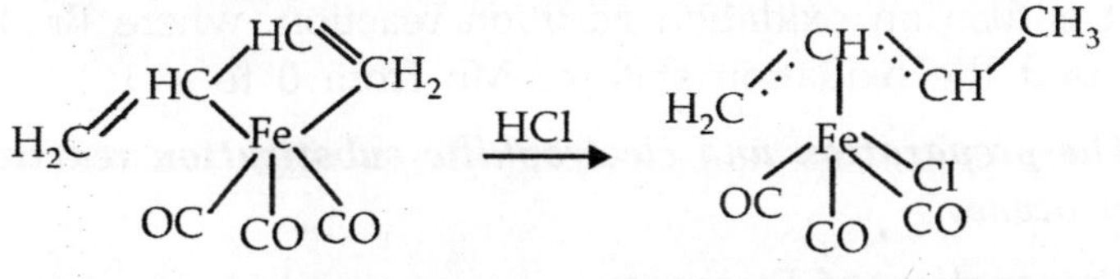

$$(\eta^5cp)_2Fe + HBF_4 \longrightarrow (\eta^5cp)_2FeH^+ + BF_4^-$$

Addition of H^+ does not involve any addition of electrons. Hence the 18 electron rule is obeyed. The structure of the product can be represented as

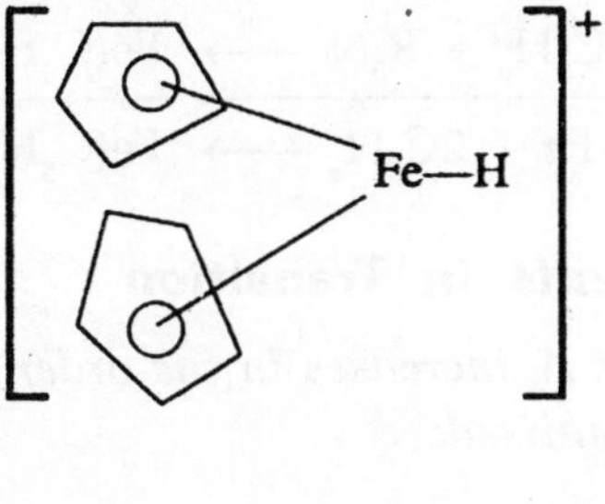

From among the following reactions, the type of reaction involved, viz., oxidative-addition, reductive-elimination, insertion or addition, identified:

(i) $[RhI_3(CO)_2CH_3]^- \longrightarrow [RhI_3CO(solvent)(COCH_3)]^-$

(ii) $Co_2(CO)_8 + H_2 \longrightarrow 2[CoH(CO)_4]$

(iii) $Mn_2(CO)_{10} + Br_2^- \longrightarrow 2MnBr(CO)_5$

(*i*) $[RhI_3(CO)_2CH_3]^- \longrightarrow [RhI_3CO(solvent)COCH_3]^-$

It is a type of insertion reaction in which solvent is inserted while methyl migrates,

(*ii*) $Co_2(CO)_8 + H_2 \longrightarrow 2[CoH(CO)_4]$

It is a type of oxidative addition reaction. In these reactions a low valent transition metal complex reacts with a molecule to give a product in which both the oxidation number and coordination number of the metal are increased. In the given reaction H has entered as the hydride ion and hence the oxidation number of cobalt has increased from 0 to + 1.

$$Mn_2(CO)_{10} + Br_2 \longrightarrow 2MnBr(CO)_5$$

It is also an oxidation addition reaction, where Br^- has increased the oxidation state of Mn from 0 to + 1.

The preparation and electrophilic substitution reactions of ferrocene.

Preparation of Ferrocene:

(i) $\underset{\substack{\text{Cyclopentadiene}\\ \text{(in excess)}}}{2C_5H_6} + 2N(Et)_2H + FeCl_2 \longrightarrow \underset{\text{Ferrocene}}{Fe(C_5H_5)_2} + 2[N(Et)_2H_2]^+Cl^-$

(ii) $Fe + R_3NH^+Cl^- \longrightarrow FeCl_2 + 2R_3N + H_2$

$$FeCl_2 + 2C_5H_6 + R_3N \longrightarrow Fe(C_5H_5)_2 + 2R_3NH + Cl^-$$

$$Fe + 2C_5H_6 \longrightarrow Fe(C_5H_5)_2 + H_2$$

Various Elements in Transition

The fact that Δ_0 increases in the order $CrCl_3^{3-}$, $Cr(NH_3)_6^{3+}$, $Cr(CN)_6^{3-}$, rationalised:

Inspire of negative charge of Cl^-, Δ_0 for Cl^- is less than Δ_0 for NH_3 because of the repulsive interaction between the *pπ* electrons of Cl^- and the *d* π electrons of Cr^{3+}. Δ_o for CN^- is very high because of back bonding *ie.,* the shift of electron density from the *dπ* orbitals of Cr^{3+} to the antibonding π* orbitals of CN^-. The latter interaction stabilises the *dπ* orbitals and increases Δ^0.

(A ligand such as H^-, NH_3 or CH_3^- can engage only in a bonding. However, a ligand such as Cl^-, O^{2-} or CO_3^{2-} has electrons in essentially non-bonding *p*-orbitals of the donor atoms which can interact with appropriate *d*-orbitals of the metal atom. Similarly, a ligand such as CN^-, CO or PR_3 has empty π-orbitals which can interact with *d*-metal orbitals. In the case of CN^- and CO, the empty π-orbitals are the antibonding π-MOs of these diatomic species.)

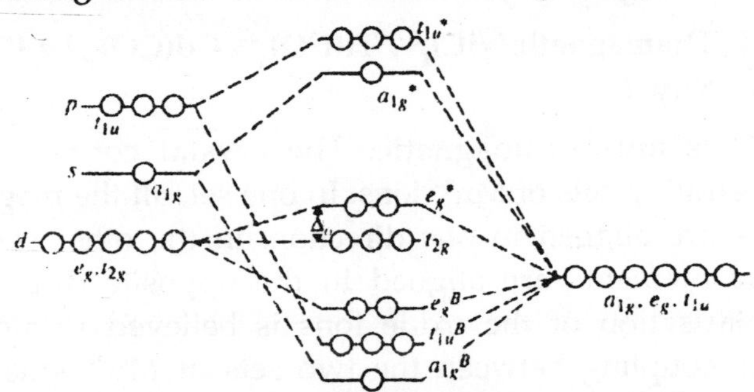

Molecular orbital energy level diagram for an octahedral complex with only o bonding.

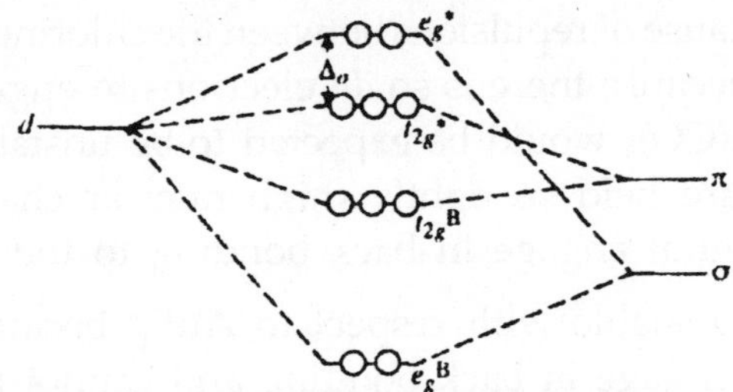

Energy level diagram showing the interaction of the $t_{2g}d$ orbitals with filled π-orbitals of the ligands. The non-bonding ligand π-orbitals are not shown. Notice that the interaction tends to decrease Δ_0.

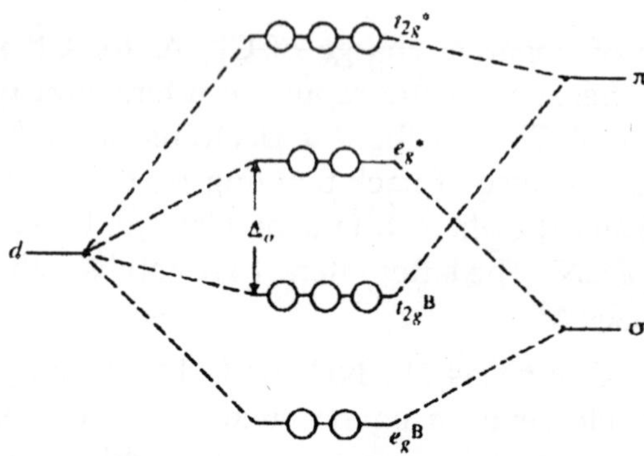

Energy level diagram showing the interaction of the $t_{2g}d$ orbitals with empty π-orbitals of the ligands. The ligand π-orbital combinations of other symmetries are not shown. Notice that the interaction tends to increase Δ_0.

An explanation for the fact that MnO (NaCl structure) has no net magnetic moment:

$ZrCl_4$, Diamagnetic $NiCl_4^{2-}$, $Ti(CO)_4^{4+}$, $Cd(CO)_3$, $Fe_3(CO)_{12}$, AuF_5^{2-} ? Why ?

MnO is antiferromagnetic. The crystal contains two interpenetrating sets of Mn^{2+} ions. In one set, all the magnetic moments are aligned in one direction, in the other set, the magnetic moments are aligned in the opposite directions. Spin polarisation of the oxide ions is believed to aid the magnetic coupling between the two sets of Mn^{2+} ions.

Diamagnetic $NiCl_4^{2-}$ could be square planar; such a structure would be unstable with respect to tetrahedral form, probably because of repulsions between the chlorines. $Ti(CO)_4^{2-}$ is unstable because there is so *dn* electrons to engage in back bonding. $Cd(CO)_3$ would be expected to be unstable because d-electrons are held so tightly (high nuclear charge of Cd) that they cannot engage in back bonding to the CO group.

AuF_5 is unstable with respect to AuF_4^- because fluoride ions cannot engage in back bonding and cannot remove the negative charge which builds up on the Au atom when five ligands are coordinated to it.

The d-orbital splitting diagram for a square pyramidal and a trigonal bipyramidal complex:

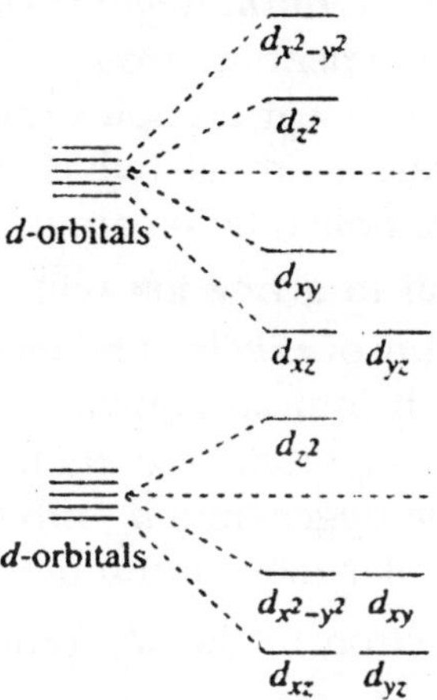

d-orbital Splitting in Square Pyramidal Complex: Square pyramidal has C_{4V} geometry.

In this pyramid base in *xy* plane and energies of various orbitals in Dq units are $d_{x^2-y^2}$ (9.14), d_{z^2} (0.86), d_{xy} (–0.86),d_{xz} and d_{yz}(– 4.57).

d-orbital Splitting in Trigonal Bipyramidal Complex: The trigonal bipyramidal has D_{3n} symmetry.

Pyramid base in *xy* plane and energies of various orbitals in Dq units are d_{z^2} (7.07), d_{x^2y} and d_{xy} (–0.82), d_{xz} and d_{yz} (–2.72).

The complex $[Fe(H_2O)_6]^{2+}$ displays two overlapping absorption bonds at –1000 nm. An explanation.

$[Fe(H_2O)_6]^{2+}$ is a high spin d^6 ion, the ground state of d^6 ion is ^5D. The ground state energy level $^5T_{2g}$ originates from the configuration $f_{2g}{}^4\, e_g{}^2$. The excited energy level ^5E_f originates from $t_{2g}{}^3\, e_g{}^3$. The absorption band observed is, therefore, assigned to $^sT_{2g} \longrightarrow {}^5E_g$. The band appears to be composite one due to splitting of excited state energy level ^5E_g into two non-degenerate energy levels due to John Tellar distortion.

The explanation of quenching of orbital angular momentum on the basis of crystal field theory:

Quenching of the Orbital Angular Momentum: As electrons will have orbital momentum around the given axis (say az-axis), if the orbital that it occupies can be transformed into an entirely equivalent orbital (which should be degenerate also with it) by a simple rotation around the axis. The electron in the $d_{x^2 \cdot y^2}$ orbital in a free ion will have a orbital angular magnetic momentum of ±2*h*/2π units about the z-axis as rotation by 45° will carry it into an equivalent d_{xy} orbital. Similarly, a 90° rotation of the d_a orbital around the z-axis will transform the orbital into the degenerate d_{yx} orbital. Hence, a d_{zx} or d_{yx} orbital has an angular momentum of ± 1 *(h/2π)* units around the z-axis. An electron in the d_{z^2} orbital will not have any orbital momentum along the z-axis as it cannot be transformed into $d_{x^2-y^2}$ or d_{zx}, etc.orbitals by simple rotation.

In an octahedral ligand field, only t_{2g} orbitals remain degenerate and rotationally related. The E_g orbitals get separated by 10 Dq.

Hence, orbital momentum due to the $d_{x^2 \cdot y^2}$ orbital electron get quenched, and the spin-only formula should apply.

It can be seen that the orbital angular momentum formula should be important for the high spin d^1, d^2, d^6 and d^7 ion complexes and low spin d^4, d^5 ions in octahedral field. In the tetrahedral field, the high spin d^3, d^4, d^8 and d^9 ions should have a significant contribution from the orbital angular momentum. The magnetic moment of $[CoCl_4]^{2-}$ [4.4 BM] and that of $[Co(H_2O)_6]^{+2}$ (5.0) confirm the above statements, the orbital moment contributes for the high spin octahedral, but not for the tetrahedral complexes.

Even for the other ions, where no orbital moment is expected, the observed values significantly depart from the spin-only formula (though the differences are small). This is attributed to the spin-orbital interactions which oppose the quenching of the orbital moments by mixing the orbitals. This mixing (or hybridisation) of orbitals, separated by energy difference Δ, changes the magnetic moment to

$$\mu_{eff} = \mu_s \left(1 - \alpha \frac{\lambda}{\Delta}\right) \xi = \pm 25\lambda.$$

α is a constant depending on the spectroscopic state and the no. of *d* electrons, being two for 2_D or 5_D; four for 3_F or 4_F and zero for 6_S. λ is the spin-orbit coupling constant for different J states and related to the spin-orbit coupling constant ξ as given (it is positive for less than half filled subshell and negative for more than half filled subshell). This explains the generally, lower μ_{eff} values obtained for Cr^{2+}, Cr^{3+}, V^{3+} and V^{+} and higher values for spin free Fe^{2+}, Co^{2+}, Ni^{2+} and Cu^{2+} complexes. Greatest deviations occur for the Co^{2+} and Fe^{2+} complexes, for which unquenched orbital moments contribute significantly.

The observed magnetic moments for the metals in t_{2g} ground state are temperature dependent and usually depart from the μ_s values due to probably the t_{2g} electron delocalisations and lower symmetry ligand field components.

The structures of: (i) Perovskite, $CaTiO_3$, (ii) Rutile, (iii) CuCI.

(i) ***Perovskite:*** The mineral perovskite, $CaTiO_3$, has a structure in which the oxide ions and the large cations (Ca^{2+}) forms a *ccp* array with the smaller cation (Ti^{4+}) occupying those octahedral holes formed exclusively by oxide ions. This structure is adopted by many ABO_3 oxide ion.

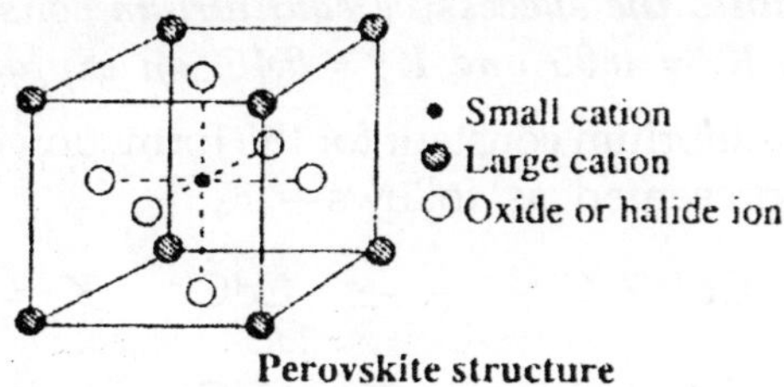

Perovskite structure

(ii) ***Rutile:*** The rutile structure, named after mineral TiO_2, is very common among oxides and fluorides of the MF_2 and MO_2 types, where the radius ratio favours coordination number 6 for the cation. Each Ti^{4+} is octahedrally surrounded by six O^{2-} ions and each O^{2-} ion has three Ti^{4+} ions round it in a plane triangular arrangement.

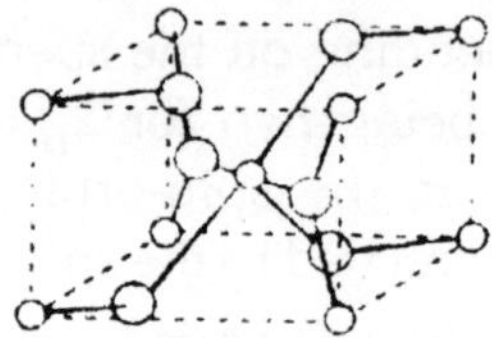

Rutile (TiO_2)

(iii) ***CuCl:*** CuCl has a covalent, tetrahedral zinc blende [ZnS] structure with the metal atom tetrahedrally surrounded by halogen atoms. Cu^+ ions occupy half of the suitable tetrahedral sites in a face centred cubic lattice formed by Cl^- ions. Thus each Cl^- ion is also surrounded tetrahedrally by four Cu^+ ion.

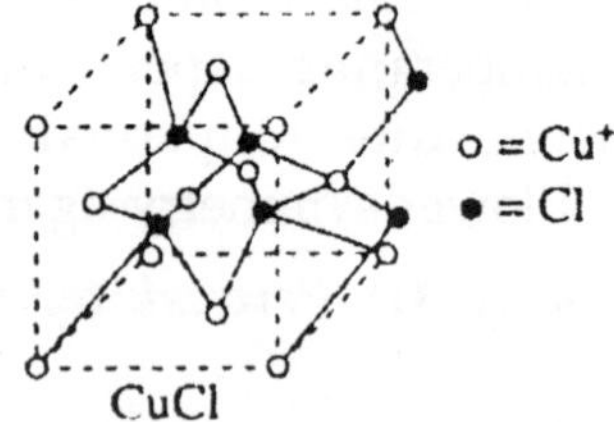

Structure of CuCl

In vapour state, CuCl is polymeric and contains a cyclic ring of alternating Cu and Cl atoms.

The formation of cadmium (II) complexes with chloride anion exhibits the successive equilibrium constant K_1 = 1.56, K_2 = 0.54, K_3 = 0.05 and K_4 = 0.46, an explanation.

The equilibrium constant for the formation of this complex can be represented as follows—

$$Cd^{2+} + Cl^- \rightleftharpoons CdCl^+ \quad K_1 = 1.56$$

$$CdCl^+ + Cl^- \rightleftharpoons CdCl_2 \quad K_2 = 0.54$$

$$CdCl_2 + Cl \rightleftharpoons CdCl_3^- \quad K_3 = 0.05$$

$$CdCl_3^- + Cl^- \rightleftharpoons CdCl_4^{2-} \quad K_4 = 0.46$$

From these data it can be seen that $CdCl_2$ is very stable with respect to disproportionation of $CdCl^+$ and $CdCl_3^-$. The reason may be an abrupt change in coordination number and hybridisation in the sequence of complexes.

10 ml of a 0.001 M metal solution and 10 ml of a 0.003 M ligand solution were mixed. The final concentration of the metal complex ML_3 was 0.000499 M. The stability constant of the complex, calculated:

The reaction can be written as

	M^{n+} +	3L	$\rightleftharpoons$	$[ML_3]^{n+}$
Initial	0.001 M	0.003 M		0M
Equilibrium	0.001-0.000499	0.003 – 3 × 0.000499		0.000499M
	= 0.000501 M	0.003 – 0.001497		
		= 0.001503		

The stability constant of the complex is given by

$$K = \frac{[ML_3]^{n+}}{[M^{n+}][L]^3} = \frac{0.000499}{0.000501 \times (0.001503)^3}$$

$$= \frac{0.000499}{0.000501 \times 3.3953 \times 10^{-9}}$$

$$= 0.29335 \times 10^9 = 2.9335 \times 10^8$$

The concentration of $[Cu(H_2O)_4]^{2+}$ (represented by $[Cu^{2+}]$) in 1.0 L of a solution made by dissolving 0.10 Cu^{2+} in 1.00 M aqueous ammonia K_s = l × 10^{12}, calculated:

$$Cu^{2+} + 4NH_3 \rightleftharpoons [Cu(NH_3)]^{2+}$$

$$K = \frac{[Cu(NH_3)_4^{2+}]}{[Cu^{2+}][NH_3]^4} = 1 \times 10^{12}$$

The reaction between Cu^{2+} and NH_3 goes almost to completion as is suggested by large magnitude of the formation constant. Thus the equilibrium concentration of tetraamine complex is approximately 0.10 M (If x mol/L Cu^{2+} ion still remain uncoordinated, the concentration of the complex will be $0.10 - x$). With the formation of 0.10 M $Cu(NH_3)_4^{2+}$, which requires 4 mol NH_3 per mol complex, the ammonia concentration becomes $1.00 - 4(0.10) = 0.60$ M (Dissociation of the complex provides $4x$ mol NH_3 per x mol Cu^{2+}).

	initial	used up	produced	equilibrium
$[Cu^{2+}]$	010	$0.10 - x$		x
$[NH_3]$	1-00	$4(0.10 - x)$		$0\text{-}60 + 4x$
$[Cu(NH_3)_4^{2+}]$	0.00		$0.10 - x$	$0.10 - x$

$$K = \frac{0.10}{x(0.60)4} = 1 \times 10^{12}$$

$$x = 8 \times 10^{-13} \text{ M}$$

The ferromagnetism and antiferromagnetism, explained:

Ferromagnetism and Antiferromagnetism: Ferromagnetism and antiferromagnetism are cooperative phenomenon and are, therefore more likely to occur where there is a high concentration of paramagnetic species. A distinction is therefore, made between magnetically concentrated and magnetically dilute materials. In magnetically dilute substances, *i.e.*, in which the metal ion containing unpaired electrons are isolated from one another by a cluster of ligands, each paramagnetic metal ion will act independent of other such ions because the distance between the metal ions is too large to allow the paramagnetic spin of one to interact with that of the other ion. But, this is not the case in magnetically concentrated substances, *i.e.*, in which the paramagnetic ions allow their total electron spins to interact with one another through exchange interaction.

If it is assumed that only the spin magnetic moment contributes towards μ_{eff} of a magnetically concentrated substances, then ΔE, the energy of spin interaction of paramagnetic ion is given by

$$\Delta E = 2J\,(S)_i(S)_k$$

where $(S)_i$ and $(S)_k$ are the total electron spin of the *ith* and £th metal ion of the substance. J is known as exchange coupling constant. J is a measure of the strength of interaction between the total electron spin of the metal ions. If J is positive, then the total electron spin (S)s of all the paramagnetic ions in the magnetically concentrated substance in ground state get aligned in the same direction. This is called ferromagnetism.

If J is negative, then total electron spins (S)s of all the paramagnetic ions in magnetically con-centrated substance in ground state are *paired up*. This is called *antiferromagnetism.*

The ferromagnetic alignments of spins produce χ greatly in excess of that produced by normal paramagnetic alignments. χ in ferromagnetic substances is dependent on the strength of magnetic field applied. The ferromagnetic behaviour exists upto curie temperature or curie point.

The total electron spin vector (S)s of atom or ions in a ferromagnetic material are all coupled parallel to one another in groups or domains. In presence of magnetic field the magnetic moments of all the domains start aligning parallel to the direction of the applied field. This tendency is opposed by thermal agitation which tends to randomise the magnetic moments of domains. However, the thermal energy is not always able to randomise the magnetic moment once these have been aligned. This results in the retention of magnetic moment of domains in a ferromagnetic substance even after the external field is removed. This is known as *hysteresis.*

Antiferromagnetic spin exchange interactions cause a lowering of χ. In most of the antiferromagnetic substances, the electron spin interactions leading to the pairing of total electron spins of neighbouring metal ions occurs through the intervening diamagnetic atom or ions. This type of total electron spin exchange occurs through linear O^{2-} or F^- bridges lying between the paramagnetic metal ions of a ferromagnetic substance.

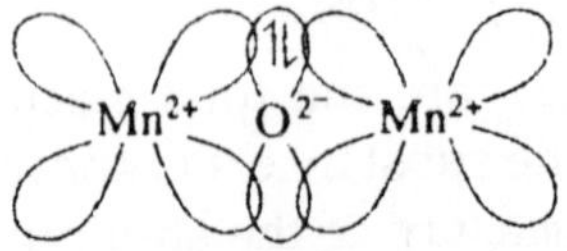

Antiferromagnetism is very much temperature dependent. For antiferromagnetic substances there is a characteristic temperature, known as *Neel temperature* or *Neel point* above which the antiferromagnetic substance behave as a normal paramagnetic substance.

The Magnetic Moments from Magnetic Susceptibilities:

When a substance is subjected to magnetic field H, and a magnetisation I is induced. The ratio I/H is called the volume susceptibility κ and can be measured by a number of techniques, including Gouy balance method, Faraday method and an *nmr* method. The volume susceptibility is related to the gram susceptibility χ and the molar susceptibility χ_m.

$$\chi = \frac{\kappa}{d} \qquad \chi_m = \frac{\kappa M}{d} \qquad \ldots(1)$$

where d and M are the density and mol. wt. of the substance. For a paramagnetic substance κ, χ and χm are positive quantities. The effective magnetic moment μ_{eff} is calculated from the relation

$$\mu_{eff} = 2.83\sqrt{\chi_M T} \qquad \ldots(2)$$

where T is the absolute temperature and the constant 2.83 is obtained from quantum mechanical calculations

$$\chi_{para} = N_A \mu^2_{eff}/3kT \qquad \ldots(3)$$

N_A = Avogadro number, k = Boltzmann constant. On substituting the values we get the equation (2).

Magnetic susceptibility would be independent of the strength of the external field applied. However, magnetic susceptibility would depend on temperature since the thermal agitation would tend to randomise the atomic or molecular magnetic dipoles which get aligned in the direction of the magnetic field applied. The magnetic susceptibility of a paramagnetic substance, χ_{para} would, therefore, decrease with increase in temperature. The variation of χ_{para} with temperature is expressed in the form of Curie law as

$$\chi_{para} = \frac{c}{T} \quad \text{where } c = \text{Curie constant}$$

The Curie law is obeyed with great accuracy by some systems such as $[FeF_6]^{3-}$. However, many paramagnetic material deviate slightly from the ideal behaviour, and obey Curie-Weiss law

$$\chi_{para} = \frac{c}{T-\theta} \quad \text{where } \theta = \text{Weiss constant}$$

The origin of the intense colour:

(i) [Cu(b py)] ***bpy*** **= 2, 2′ = bipyridine (ii) $K_3[RuCl_6]$**

(i) *[Cu (bpy)]:* bpy or 2, 2′ bipyridine is a π-acceptor ligand. The origin of intense colour of the complex is Metal → Ligand charge transfer spectra as the ligands have empty π-antibonding orbitals. Thus the bonds responsible are mainly $d\pi^*$ charge transfer bands. Charge transfer processes from metal to ligand are favoured in complexes that have occupied metal-centred orbitals and vacant low lying ligand-centred orbitals.

(ii) *$K_3[RuCl_6]$:* The origin of the intense colour in this complex is Ligand → Metal charge transfer spectra charge transfer transitions $t''_{1u}, {}''t_{2u} \rightarrow t^*_{2g}$ occur along with other charge transfer transition $t''_{1u}, t''_{2u} \rightarrow e_g$.

A charge transfer process is similar to redox process. The energy of transition measures the tendency of the redox reaction to proceed. As the charge transfer bands are neither multiplicity forbidden, nor Laporte forbidden, they have high absorption intensities.

The unusual positive oxidation states of Mn:

Manganese shows a number of oxidation states due to negligible energy difference between *ns* and $(n-1)d$ orbitals, so electrons from both the orbitals can participate in chemical reaction. The outer electronic configuration of Mn is $3d^5\ 4s^2$. The most stable oxidation states of Mn are + **II**, + IV and +VII. The unusual positive oxidation states shown by Mn are + I, + III, + IV and + VI. Other than this Mn shows oxidation states 0, –1, –II and –III.

Mn shows (– III) oxidation in the compound $[Mn(NO)_3CO]$. The examples of compounds in various oxidation are given below in table—

Oxidation States and Sterochemistry of Manganese

Oxidation state	*Coordination number*	*Geometry*	*Examples*
Mn^{-III}	4	Tetrahedral	$Mn(NO)_3CO$
Mn^{-II}	4 or 6	Square	$[Mn(phthalocyanine)]^{2-}$
Mn^{-I}	5	*tbp*	$Mn(CO)_5^-$, $[Mn(CO)_4PR_3]^-$
	4 or 6	Square	$[Mn\ (phthalocyanine)]^-$
Mn^o	6	Octahedral	$Mn_2(CO)_{10}$
Mn^I, d^6	6	Octahedral	$Mn(CO)_5Cl$, $K_5[Mn(CN)_6]$, $[Mn(CNR)_6]+$
Mn^{II}, d^5	2	Linear	$Mn[C(SiMe_3)_2]_2$
	4	Tetrahedral	$MnCl_4^{2-}$, $MnBr_2(OPR_3)_2$, $[Mn(CH_2SiMe_3)_2]_n$
	4	Square	$[Mn(H_2O)_4]SO_4.H_2O$, $Mn(S_2CNE_2)_2$
	5	Distorted *tbp*	$MnBr_2[(MeHN)_2CO]_3$
	6	*tbp* Octahedral	$[Mn(trenMe_6)Br]Br$
	7	NbF_7^{2-} structure	$[Mn(H_2O)_6]^2+$, $[Mn(NCS)_6]^{4-}$
		Pentagonal	$[Mn(EDTA)H_2O)^{2-\ d}$
		bipyramidal	$MnX_2(N_5$ macrocycle)
	8	Dodecahedral	$(Ph_4As)_2Mn(NO_3)_4$
Mn^{III}, d^4	4	Square	$[Mn(S_2C_6H_3Me)_2]^-$
	5	*sp*	MnX sal_2en, $[bipyH_2]MnCl_5$
	5	*tbp*	$MnI_3(PMe_3)_2{}^f$
	6	Octahedral	$Mn(acac)_3$, $[Mn(ox)_3]^{3-}$, MnF_3(distorted),
	7		$Mn(S_2CNR_2)_3$
			$[Mn(EDTA)H_2O]^-$, $MnH_3(dmpe)_2$

Mn^{IV}, d^3	4	Tetrahedral	Mn (l-norbornyl)$_4$
	6	Octahedral	MnO_2, $MnMe_4$(dmpe), $MnCl_6^{2-}$ - $Mn(S_2CNR_2)_3$
Mn^{V}, d^2	4	Tetrahedral	MnO_4^{3-}
Mn^{VI}, d^1	4	Tetrahedral	MnO_4^{2-}
Mn^{VII}, d^0	3	Planar	MnO_3^{+}
	4	Tetrahedral	MnO_4^{-}, MnO_3F

Zinc sulphide forms a cubic unit cell of length 6×10^{-10} m. Zinc ions form a face centred cubic lattice and sulphide ions occupy the centre of the alternate small cubes:

(a) Find the position of zinc and sulphide ions,

(b) Assign the coordinates,

(c) Effective number of Zn and S ions and formula of zinc sulphide,

(d) Density of ZnS if the atomic weight of Zn and S are 65.4 and 32 g mol^{-1},

(e) Smallest distance between zinc and sulphide ion,

(f) Coordination number of each ion.

The structure of zinc sulphide crystal is face centred cubic lattice.

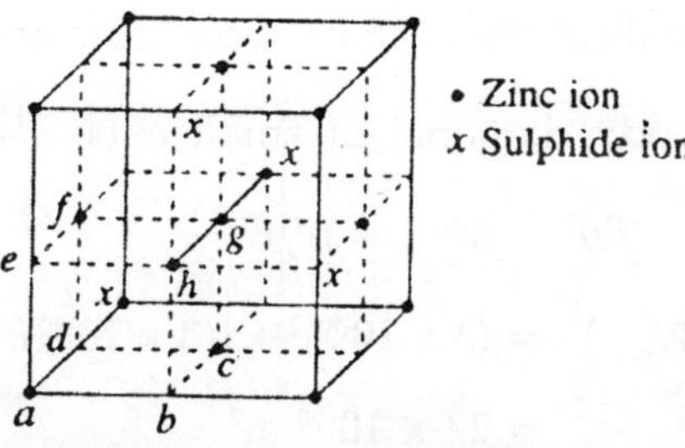

(a) The • represents zinc ion and x represents sulphide ions.

(b) The cordinates of zinc ions are

$$(0, 0, 0), \left(\frac{1}{2}, \frac{1}{2}, 0\right), \left(\frac{1}{2}, 0, \frac{1}{2}\right), \left(0, \frac{1}{2}, \frac{1}{2}\right)$$

For sulphide ions are

$$\left(\frac{1}{4}, \frac{1}{4}, \frac{1}{4}\right), \left(\frac{1}{4}, \frac{3}{4}, \frac{3}{4}\right), \left(\frac{3}{4}, \frac{1}{4}, \frac{3}{4}\right), \left(\frac{3}{4}, \frac{3}{4}, \frac{1}{4}\right)$$

(c) Effective number of zinc ions in unit cell are

$$8\times\frac{1}{8}+6\times\frac{1}{2} = 4$$

Effective number of sulphide ions are 4 × 1 = 4

So the molecular formula is ZnS.

(d) Density is given by the formula where,

$$P = \frac{Z\times M}{a^3\times N_0}$$

where, Z = Number of particles per unit cell

M = Atomic mass in kg mol^{-1}

a = Edge of the unit cell

N_0 = Avogadro number

$$= \frac{(4)(65.4+32)\times 10^{-3}\text{ kg mol}^{-1}}{(6.023\times 10^{23}\text{ mole mol}^{-1})(6\times 10^{-10})^3\text{ m}^3}$$

$$= 2.995 \times 10^3 \text{ kg m}^3$$

(e) The smallest distance between zinc and sulphide ions is $\frac{1}{2}$ of the body diagonal of small cube *abcdefgh*

$$(fb)^2 = (db)^2 + (df)^2$$

$$\left(R_{Zn^2}+2R_{S^{2-}}+R_{Zn^{2+}}\right)^2 = (3\times 10^{-10})^2 + [(3\times 10^{-10})^2 + (3 + 10^{-10})^2]$$

$$= 27 \times 10^{-20}\text{ m}^2$$

$$2\left(R_{Zn^2}+2R_{S^{2-}}\right) = 3\sqrt{3}\times 10^{-10} = 3 \times 1.732 \times 10^{-10}\text{ m}$$

$$R_{Zn^2}+R_{S^{2-}} = \frac{5.196\times 10^{-10}}{2} = 2.598 \times 10^{-10}\text{ m}$$

, *i.e.* smallest distance between zinc and sulphide ions is 2.598×10^{-10} m.

(f) Each sulphide ion has four zinc as the nearest neighbours arrange tetrahedrally and each zinc ion is surrounded by four sulphide ions tetrahedrally.

The Chemistry of Isopoly and Heteropoly Acids/salts of Molybdenum and Tungsten:

The poly acids of molybdenum and tungsten are of two types—

1. The isopoly acids and their related apions which contain only molybdenum or tungsten along with oxygen or hydrogen;
2. The heteropoly acids and anion which contain one or two atoms of another elements in addition to molybdenum, tungsten, oxygen and hydrogen.

The polyanions are built primarily of MO_6 octahedra, but they are prepared by starting with MO_4^{2-} ions. The polymolybdates are formed by the condensation of MoO_4^{2-} unit and their composition depends upon the pH of the solution.

$$Mo_4O^{2-} \xrightarrow{pH\,6} [Mo_7O_{24}]^{6-} \xrightarrow{pH\,2} [Mo_8O_{26}]^{4-}$$

The formation of isopolytungstate is similar, the simple tungstate WO_4^{2-} exists in strongly basic solution. Acidification results in the formation of polymers built up from WO_6 octahedra. The important species formed are $W_7O_{24}^{6-}$ and $W_{12}O_{42}^{12-}$.

Heteropoly acids and their salts are formed when molybdate and tungstate solutions containing other oxoanions (*e.g.*, PO_4^{3-} and SiO_4^{4-}) or metal ions are acidified. At least 35 elements are known to be capable of functioning as the hetero atoms.

$$HPO_4^{2-} + MoO_4^{2-} \xrightarrow{H^+,\,25°C} [PMo_{12}O_{40}]^{3-}$$

$$WO_4^{2-} \xrightarrow{H^+\text{ to pH}=6} \xrightarrow{Co^{2+},\,100°} [Co_2W_{11}O_{40}H_2]^{8-}$$

Some Representative Heteropoly Salts and their Nomenclature

Formula	*IUPAC Names*
$Na_3[P^vMo_{12}04o]$	Sodium 12-molybdophosphate; sodium dodecamolybdophosphate
$H_3[P^vMo_{[2}04o]$	12-Molybdophosphoric acid; dodecamolybdophosphoric acid
$K_3[Co_2{}^{II}W_{12}O_{42}]$	Dimeric potassium 6-tungstocobaltate; dimeric potassium hexatungstocobaltate (II)
$Na_3[Ce^{IV}Mo_{12}C_{42}]$	Sodium 12-molybdocerate (IV); sodium dodecamolybdocerate (IV)

In the structure of isopoly and heteropoly anions the tungsten and molybdenum atoms lie at the centres of octahedra of oxygen atoms and the structures are built up of these octahedra by means of shared corners and shared edges (but not shared faces).

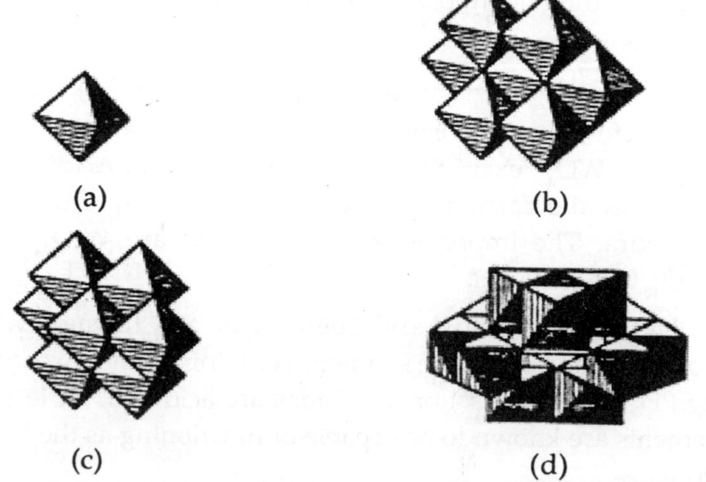

(a) (b) (c) (d)

(a) Diagrammatic representation of MoO_6 and WO_6 octahedra used in showing structures of some isopoly and heteropoly anions.

(b) The structure of the para-molybdate anion, $[Mo_7O_{24}]^{6-}$.

(c) **The structure of the octamolybdate anion, $[Mo_8O_{26}]^{4-}$ (note that one MoO_6 octa-hedron is completely hidden by the seven that are shown).**

(d) **The structure of the $[W_{12}O_{42}]^{12-}$ unit in the paratungstate ion.**

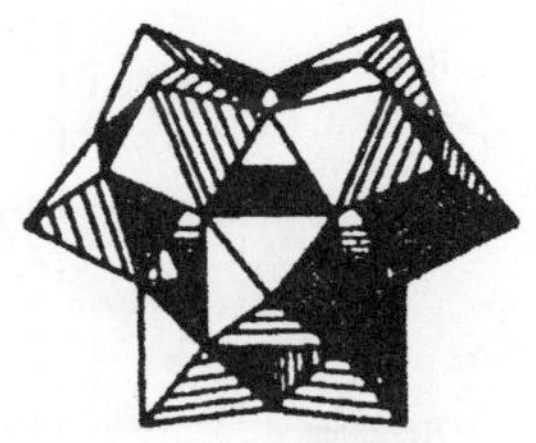

The structure of the series A 12-molybdo and 12-tungs o-heteropoly anions of general formula $[X^{n+}Mo_{12}O_{40}]^{(8-n)-}$.

The Spin Orbit Coupling for p^2 System:

Spin Orbit Coupling: When several electrons are present in a sub-shell, the overall effect of the individual orbital angular momenta l is given by resultant quantum number L, and the overall effect of the individual spin m_s is given by the resultant spin quantum number S. In an atom, the magnetic effects of L and S may interact or couple, giving a new quantum number J called total angular momen-tum quantum number which results from vectorial combination of L and S (Russell - Saunders or LS coupling).

Spin Orbit Coupling of p^2: With a p^2 arrangement, resultant orbital quantum number values of L = 2, 1 and 0, and also resultant quantum number values S = 1 and 0 are obtained. These may be coupled to give the total angular quantum number J. Each of these arrangement corresponds to an electronic arrangement called a spectroscopic state.

Each of these arrangements corresponds to an electronic arrangement sometimes called a spec-troscopic state, which is describe by a full term symbol. The letter D indicates that the L quantum number has a value of two, P indicates that L = 1 and S denotes a value of L = 0 as described previously. The lower right hand subscript denotes the value of the total quantum number J, and the upper left superscript indicates the multiplicity, which has the value of 2S + 1 (where S is the resultant spin quantum number).

1 ↑ Resultant 2 ↑ J = 3
Full Spectroscopic term symbol $^3D_3$
S = 1 Resultant L = 2 J = 2 $^3D_2$
L = 2 S = 1 Resultant J = 1 $^3D_1$
L = 2 S = 0 Resultant J = 2 $^1D_2$

1 ↑ Resultant 2 ↑ J = 2
Full Spectroscopic term symbol $^3P_2$
S = 1 Resultant L = 1 J = 1 $^3P_1$
L = 1 S = 1 Resultant J = 0 $^3P_0$
L = 1 S = 0 Resultant J = 1 $^1P_1$

0 S = 1 Resultant J = 1
Full Spectroscopic term symbol $^3S_1$
L = 0 S = 0 Resultant J = 0 $^1S_0$

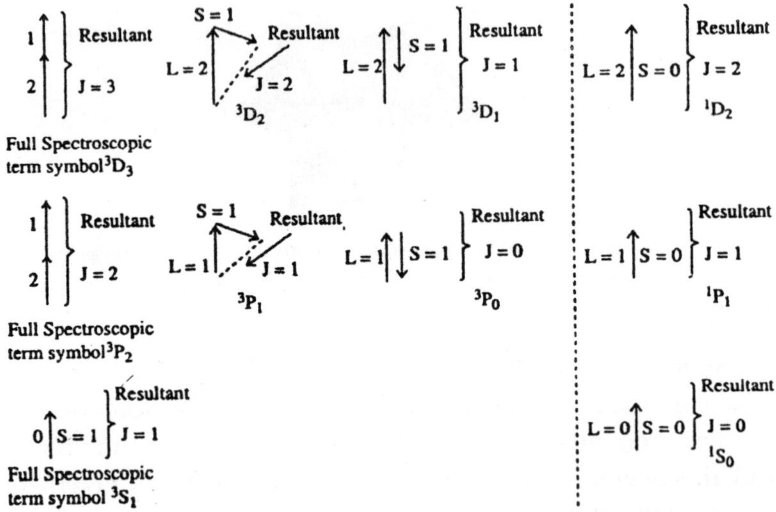

The relation between the number of unpaired electrons, the resultant spin quantum number S, and the multiplicity is given in table.

Unpaired electrons	S	Multiplicity	Name of state
0	0	1	Singlet
1	$\frac{1}{2}$	2	Doublet
2	1	3	Triplet
3	$1\frac{1}{2}$	4	Quartet
4	2	5	Quintet

Thus, the symbol $^3D_2$ (pronounced triplet D two) indicates a D state, hence L = 2; the multiplicity is three, hence S = 1 and the number of unpaired electrons is 2; and the total quantum number J = 2.

All of the spectroscopic terms derived above for a p^2 configuration would occur for an excited state of carbon Is^2, $2s^2$, $2p^1$, $3p^1$. However, in the ground state of the atom Is^2, $2s^2$, $2p^2$ the number of states is limited by the Pauli exclusion principle since no two electrons in the same atom can have all four quantum numbers the same. In the ground state configuration, the two *p* electrons both have the same values of $n = 2$ and $l = 1$ so they must differ in at least one of the remaining quantum numbers *m* or m_s. This restriction reduces the number of terms from ^3D, ^3P, ^3S, ^1D, ^1P and ^1S to ^1D, ^3P and ^1S.

This can be shown by writing down only those electronic arrangements of *m* and m_s which do not violate the Pauli exclusion principle. For *p* electrons, the subsidiary quantum number $l = 1$, and the magnetic quantum number *m* may have values from $+ l \rightarrow 0 \rightarrow l$, giving in this case values of $m = + 1, 0$ and $- 1$. There are 15 possible combinations.

The rules for determining the order of energy for different terms:

The rules are as under—

1. The terms are arranged depending upon their spin multiplicities, *e.g.*, their S-values. The most stable state has the highest S-value and stability decreases as the value of S decreases. This means that the most stable state (ground state) has maximum unpaired electrons because, this gives the minimum electrostatic repulsion. This is in accordance with Hund's rule.
2. For a given value of S, the state with the highest value of L is the most stable state. This means that if two or more terms have the same value of S (same spin multiplicity), the state with highest value of L will have the lowest energy.
3. For a given value of S and L (*i.e.*, having same S and L values), the terms with smallest J value is most

stable if the subshell is less than half filled and the term with maximum J value is most stable if the subshell is more than half filled.

Using crystal field model, explained why Fe_3O_4 has an inverse spinel structure while Mn_3O_4 has a normal spinal structure:

Normal spinels have the formula $A^{II}B_2^{III}O_4$ where A^{II} can be a group IIA metal or a transition metal in the +2 oxidation state and B^{III} is a group III A metal or a transition metal in the +3 oxidation state. The oxide ions form a close packed cubic lattice with eight tetrahedral holes and four octahedral holes per molecule of AB_2O_4. Mn_3O_4 has normal spinel structure as it is a mixed oxide of the formula $Mn^{II}(Mn_2^{III})O_4$ corresponding to the formula AB_2O_4. A^{II} ions occupy of the tetrahedral holes and B^{III} ions occupy one half of the available octahedral holes.

In the inverse spinel structure the A^{II} ions and one-half of the B^{III} ions have exchanged place that is the A^{II} ions occupy octahedral holes along with one-half of the B^{III} ions while half of the B^{III} ions are in tetrahedral holes. Fe_3O_4 (magnetite) is an example of an inverse spinel structure. Although both A and B ions in this case are iron, some are ferrous and others ferric—$Fe^{III}[Fe^{II}Fe^{III}]O_4$.

CFT helps in predicting the structures of spinels. CFSE values in octahedral and tetrahedral fields have been used for interpretation. For this it is assumed that the oxide ion O^{2-}, like water molecule produces weak field. CFSE values (in terms of Δ_0) for Mn^{3+} *(d^4)*, Fe^{3+} *(d^5)*, Mn^{2+} *(d^5)* and Fe^{2+} *(d^6)* in octahedral and tetrahedral weak field (high spin) are—

	Mn^{3+} (d^4)	Mn^{2+} (d^5)	Fe^{3+} (d^5)	Fe^{2+} (d^6)
CFSE (Oh weak field)	0.60	0	0	0.40
CFSE (Th weak field)	0.18	0	0	0.27

For Mn^{3+} *(d^4)* and Fe^{2+} *(d^6)* ions the CFSE values are greater for octahedral than tetrahedral sites. Thus Mn^{3+} and

Fe^{2+} ions will preferentially occupy the octahedral sites, maximising the CFSE values of the system. Hence in Mn_3O_4 all the Mn^{3+} ions occupy octahedral sites and all Mn^{2+} ions are in the tetrahedral sites, *i.e.*, normal spinel structure $Mn^{II}[Mn_2^{III}]O_4$. In Fe_3O_4 all the Fe^{2+} ions and half of the Fe^{3+} ions are in the octahedral sites, while the remaining half of Fe^{3+} ions occupy tetrahedral sites. Thus, it is an inverse spinel structure $Fe^{II}[Fe^{II}Fe^{III}]O_4$.

The Spectrum of Nickel(II) Octahedral Complex Ion:

Ni(II) in octahedral complexes such as $[Ni(H_2O)_6]^{2+}$ has d^8 configuration. Their spectra with three spin allowed transition are shown in Fig.

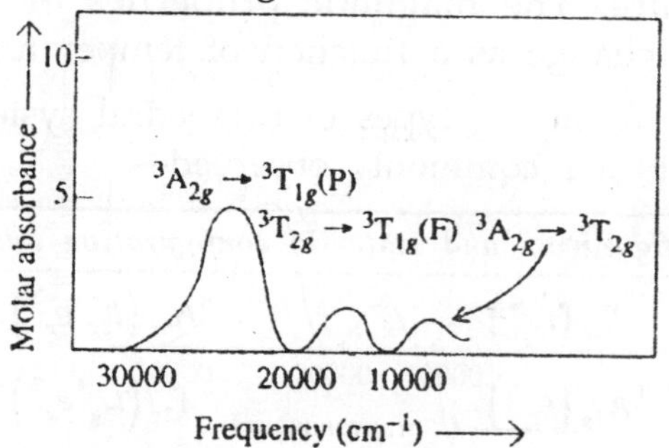

Absorption spectrum of $[Ni(H_2O_6)]^{2+}$ ion.

$$^3A_{2g} \rightarrow {}^3T_{2g} \quad 9000\ cm^{-1}$$
$$^3A_{2g} \rightarrow {}^3T_{1g}\ (F) \quad 14,000\ cm^{-1} \quad \text{for } [Ni\ (H_2O)_6]^{2+}$$
$$^3A_{2g} \rightarrow {}^3T_{1g}(P) \quad 25,000\ cm^{-1}$$

The bands are quite weak as expected. The absorption spectrum is shown in Fig.

β-elimination in transition metal alkyls ? Explained:

In higher alkyl organometallic compounds, the hydrogen present on carbon atom at β-position to the metal is labile. It brings about decomposition into metal hydride and an alkene. This is called β-elimination. The β-elimination is responsible for the instability of alkyl organometallic compounds. Therefore, the blockage of β-elimination reaction can bring about remarkably stability in metal alkyls. Therefore, the alkyls which do not contain replaceable hydrogen atom on β-carbon are highly stable, for example,

$$M—CH_2—CH_3, M—CH_2—Ar, M—CH_2—NR_2, \text{etc.}$$

High spin-low spin cross overs. The prerequisities for a coordination compound to be amenable to such cross overs are:

Spin-state Cross Overs: In the majority of complexes the ligand field is either weak enough to conclude that a high spin (HS) ground state prevails at all associable temperatures or strong enough to assure that a low spin (LS) ground state exists at all acessible temperatures. There are, however, cases where a high spin and a low-spin states are separated by only about the thermal energy prevailing at or below room temperature. The magnetic properties of the complex, therefore, change as a function of temperature.

In the following types of octahedral systems spin-state cross overs are commonly observed—

d^n	*LS configuration and state*	*HS configuration and state*	*Example*
d^5	$^2T_{2g}\left(t_{2g}^{\ 5}\right)$	$^6A_{1g}\left(t_{2g}^{\ 3}e_g^{\ 3}\right)$	Fe^{III}
d^6	$^1A_{1g}\left(t_{2g}^{\ 6}\right)$	$^3T_{2g}\left(t_{2g}^{\ 4}e_g^{\ 2}\right)$	Fe^{II}, Co^{III}
d^7	$^2E_g\left(t_{2g}^{\ 6}e_g^{\ 1}\right)$	$^4T_{1g}\left(t_{2g}^{\ 5}e_g^{\ 2}\right)$	Co^{II}

The origin of spin state cross overs is related to magnitude of the orbital splitting (Δ_0) and the spin-pairing energy P. When P and Δ_0 are about equal, the high spins and low spin states will have very similar energies. Because of the change in M-L bond lengths, which decrease in the LS state (by 0.1 – 0.2 Å) the behaviour of the system is strongly dependent on the pressure as well as temperature.

HS complexes are associated with the condition $\Delta_0 < P$ and low spin complexes with $\Delta_0 > P$. For complexes in which the energy difference between A_0 and P is relatively small, an intermediate field situation is possible for the two spin states to coexist in equilibrium with each other. Consider the Fe^{2+} ion. At the two extremes, it forms high spin paramagnetic

$[Fe(H_2O)_6]^{2+}$ (S = 2) and low spin diamagnetic $[Fe(CN)_6]^{4-}$ (S = 0). The Tanabe-Sugano diagram pertaining to these d^6 complexes shows that near the cross over point between weak and strong fields the difference in energy between the spin-free ($^5T_{2g}$ and spin paired (^1A_g) ground states becomes very small.

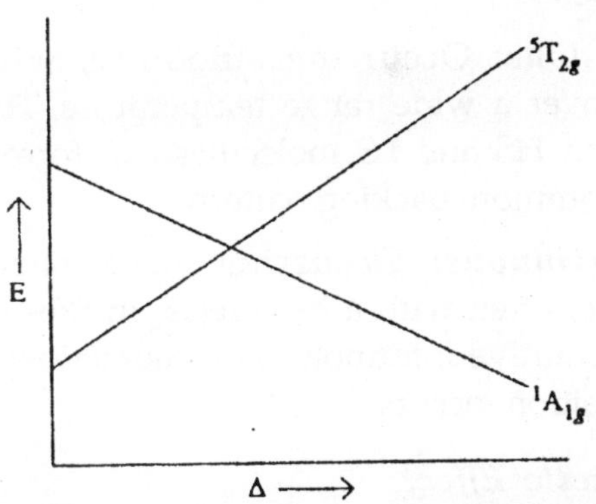

Variation in energies of the $^5T_{2g}$ and $^1A_{1g}$ terms with increasing Δ_0 for d^6 octahedral complexes. At weak fields (high spin complexes) the ground term is $^5T_{2g}$, while at strong fields (low spin complexes) it is $^1A_{1g}$. Note that in the region immediately on each side of the spin cross over point, the energy difference between the two terms is small; thus high and low spin complexes may coexist.

Within this region, it is reasonable to expect that both spin states may be present simultaneously and the degree to which each is represented will depend on temperature Δ_0 - P = *k*T. For complex $[Fe(phen)_2(NCS)_2]$ illustrating these effects is shown in Fig.

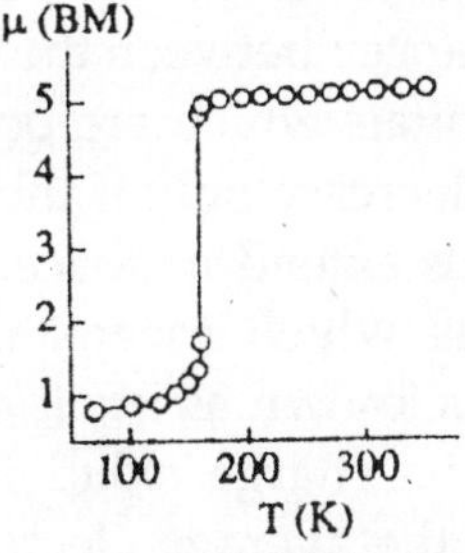

Variation in magnetic moment of $[Fe(phen)_2(NCS)_2]$ with temperature.

At high temperatures a moment coexistent with four unpaired electrons is observed, but as the temperature is decreased, a sharp drop in magnitude is observed at 175 K where the low spin form becomes dominant.

In a broader sense the spin cross-over phenomenon are of two types—

Continuous: Occurring smoothly, with little or no hysteresis over a wide range temperature. This behaviour is shown when HS and LS molecules can form solid solutions within a common packing pattern.

Discontinuous: Occurring in a narrow range of temperature, often with a hysteresis. In these cases there are strong cooperative intermolecular interactions and first-order phase transition occurs.

Nephelauxetic Effect

Electrons in the partly filled *d*-orbitals of a metal ion repel one another and give rise to a number of energy levels depending upon the arrangement of these electrons in the *d*-orbitals. The energy of each of these levels can be expressed in terms of some inter-electronic repulsion parameters. It is observed experimentally that magnitude of these inter-electronic repulsion parameters always decreases on the complexation of the metal ion. This is possible only if the magnitude of inter-electronic repulsion between the d-electrons of the metal ion decreases on complexation.

The magnitude of inter-electronic repulsion is inversely proportional to the distance *r* between the regions of maximum charge density of d-orbitals which are occupied by electrons. This repulsion would decrease only if the distance *r* increases or the lobes of d-orbitals extend in space. The extention ofthe lobes of the af-orbitals which means the expansion of d-electron charge cloud is known as *nephelauxetic effect*. Thus in the complex the electron charge cloud is more diffuse than in the free ion, hence, the average electron-electron distance is greater in the complex.

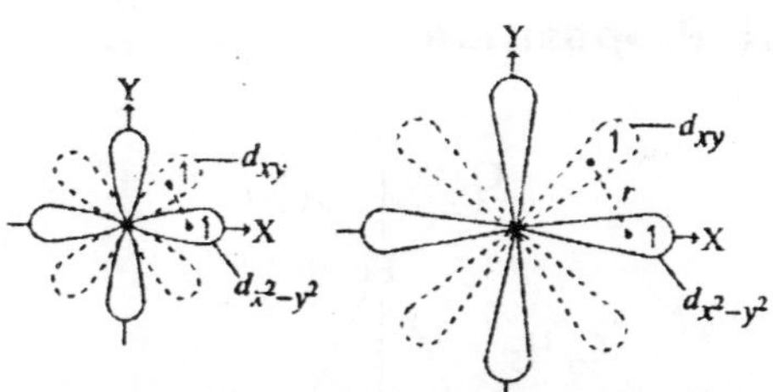

The *d*-orbitals of free metal ion. The extended *d*-orbitals of complexed metal ion.

The extension of the *d*-orbitals in space in complexed metal ion.

The extension of d-orbitals in complex occurs to maximise the overlap of ligand and metal orbitals, which is an essential condition of for covalent bonding. Thus, it is the covalent bonding which decreases the inter-electronic repulsions parameters when a free metal ion gets complexed. The larger the decrease in inter-electronic repulsion, the greater is the extent of covalent bonding in complex. The capability of *d*-orbitals of different metal ions to extend themselves forms the basis of nephelauxetic effect.

Nephelauxetic effect for a fixed ligand and different metal ions is—

Pt (IV) ~ Mn (IV) > Co (III) > Rh (III) ~ Ir (III) ~ Fe (III) > Cr (III) > Mo (III) > Ni (II) > V (II) > Mn (II).

For a given metal ion and different ligands the nephelauxetic effect is—

$(C_2H_5O)_2PSe_2^- > (C_2H_5O)_2PS_2^- > I^- > Br^- > CN^- \sim Cl^- > C_2O_4^{2-} \sim$ en $> NH_3 > (NH_2)_2CO > H_2O > F^-$

Since the inter-electronic repulsion parameters almost always decrease on complexation, the metal-ligand bonds in all the complexes must contain some contribution from covalent bonding. The most electronegative metal ions, Pt (IV) and Mn (IV) would form the most covalent bond with the least electronegative ligands.

The ratio of a given inter-electronic repulsion parameter for the metal ion in a complex to its value in the gaseous ion is expressed as nephelauxetic ratio (β).

Infrastructural Proposition

$$\begin{array}{c} (CO)_3Fe \\ C_6H_5-C \quad C-C_6H_5 \\ Fe(CO)_3 \quad Fe(CO)_3 \end{array}$$

(*i*) *Treatment of Cr(CO)*$_6$ *with LiCH*$_3$*, followed by [(CH*$_3$*)*$_3$*O]BF*$_4$*, yields the carbene complex* $(CO)_5\,Cr\,C\begin{matrix} CH_3 \\ CH_3 \end{matrix}$.

Propose a mechanism for this synthesis:

(ii) Ethylene can oxidised by aqueous solution of $PdCl_4^{2-}$:

$$C_2H_4 + PdCl_4^{2-} + H_2O \longrightarrow Pd + 2H^+ + 4Cl^- + CH_3CHO$$

The rate law is

$$\text{Rate} = \frac{k[pdCl_4^{2-}][C_2H_4]}{[H^+][Cl^-]^2}$$

Propose a Plausible Mechanism for the Reaction:

(*i*) $(OC)_5Cr = C = O + LiCH_3$

$$(OC)_5 = C(CH_3)-O^-Li^+$$

$$\updownarrow (CH)_3O$$

$$(OC)_5Cr = C\begin{matrix} OCH_3 \\ OCH_3 \end{matrix}$$

(*ii*) $PdCl_4^{2-} + C_2H_4 + H_2O \longrightarrow PdCl_2(C_2H_4)OH^- + 2Cl^- + H^+$

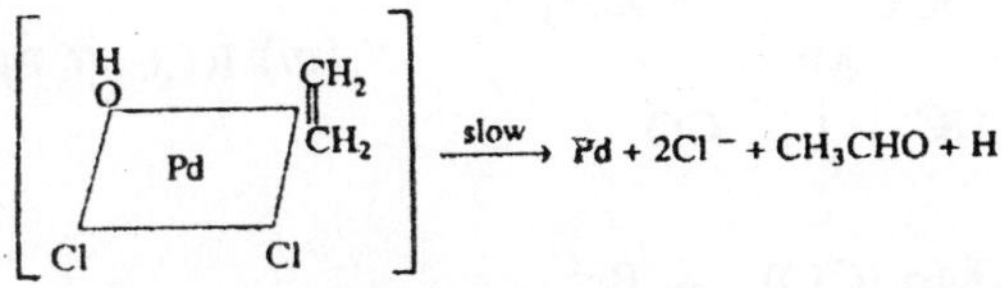

$$\left[\begin{array}{c} \text{H}_2\text{O} \cdots \text{CH}_2{=}\text{CH}_2 \\ \text{Pd} \\ \text{Cl} \quad \text{Cl} \end{array} \right] \xrightarrow{\text{slow}} Pd + 2Cl^- + CH_3CHO + H^-$$

The slow step may involve the formation of an unstable intermediate of the type $C1_2(H_2O)\ Pd—CH_2—CH_2OH^-$.

Reasonable product for the following reactions, predicted:

(i) $Cr(CO)_6$ + (butadiene) $\longrightarrow$

(ii) $Co(CO_3)NO + PPh_3 \xrightarrow{h\nu}$

(iii) $Mo(CO)_6 + CH_3CN \longrightarrow$

(iv) $Cr(CO)_6 \xrightarrow{\text{benzene}}$

(v) 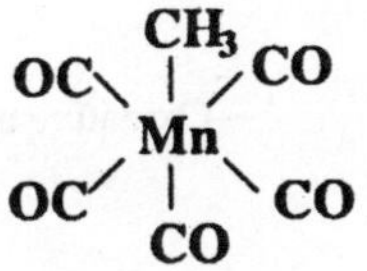 $Mn(CH_3)(CO)_5$ + $^{14}CO \longrightarrow$

(vi) $RC(=O)X + [Co(CO)_4]^- \longrightarrow$

(vii) $Mn(CO)_5Br + [Mn(CO)_5]^-$

(*i*) (butadiene)—$Cr(CO)_4$ (*ii*) $Co(CO)_2(NO)(PPh_3)$

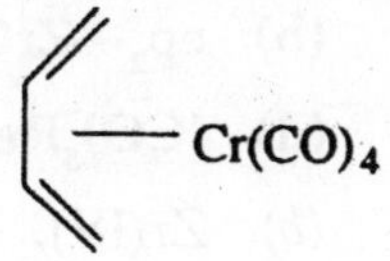

(*iii*) $Mo(CO)_3(CH_3CN)_3$ (*iv*) (benzene)—$Cr(CO)_3$

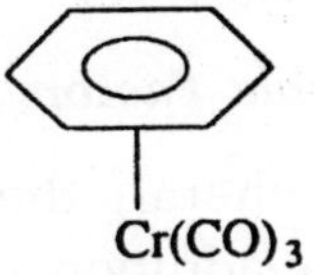

(v) Mn complex: OC, OC, CO (cis), ^{14}CO (top), $C(=O)-CH_3$ and CO (bottom) bonded to Mn

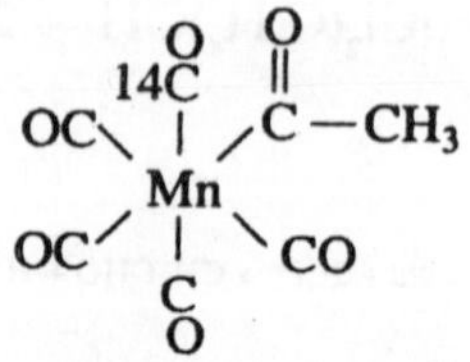

(vi) $RC(=O)Co(CP)_4 + X^-$

(vii) $Mn_2(CO)_{10}$ + Br^-

The product and state which reactions are oxidative addition and which are insertion reaction, predicted:

(i) $cp(CO)_3Re + Br_2 \longrightarrow$

(ii) trans-$(PPh_3)_2Ir(CO)Cl + H_2 \longrightarrow$

(iii) cp $(CO)_2Fe—CH_3 + PPh_3 \longrightarrow$

(iv) $Et—Re(CO)_5 + CH_3CN \longrightarrow$

(*i*) $cp(CO)_2ReBr_2$ + CO — Oxidative addition

Re (I) $\rightarrow$ Re (III)

(*ii*) $Cl(PPh_3)Ir(CO)(PPh_3)$ $\xrightarrow{H_2}$ $Ir(Cl)(OC)(PPh_3)_2(H)(H)$ —Oxidative addition

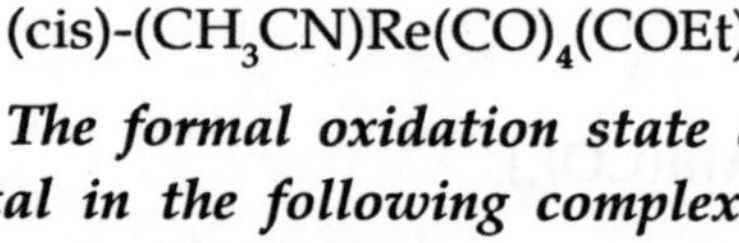

(*iii*) *cp* $(CO)(PPh_3)Fe—COCH_3$ — Insertion reaction

(*iv*) (cis)-$(CH_3CN)Re(CO)_4(COEt)$ — Insertion reaction

The formal oxidation state and d-electron count of the metal in the following complexes—

(a) $(\eta^6—C_6H_6)_2Mo$, **(b)** $cp_2—ZrCl$ (OMe)

(c) $[(PMe)_3Pd(\eta^3—C_3H_3)]^+$, **(d)** $(CO)_5Re—Et$

(*a*) Mo(O), d^6; (*b*) Zr(IV), d^0;

(*c*) Pd(II), d^8; (*d*) Re(I), d^6

Favourable Factors for the Oxidative-addition Reactions

In general, the tendency of a complex to undergo oxidative-addition reactions is governed by the amount of electron-density at the metal, *i.e.*, by the ease with which the metal can be oxidised. Thus, several general trends have been observed—

(*i*) The presence of electron rich ligarids in the coordination sphere of the metal increases the rate of oxidative-addition.

(*ii*) A low initial oxidation state of the metal is more favourable for oxidative-addition to occur *e.g.*, all other factors being equal, Fe (0) is easier to oxidise than Co (I), which is easier to oxidise than Ni (II), even though the d-electron configuration is same for these metals.

(*iii*) The tendency to oxidative-addition to occur increases down a given group *e.g.*, Ir (II) is easier to oxidise than Rh (II), which is easier to oxidise than Co (II).

(*iv*) Oxidative-addition occurs more readily in coordinatively unsaturated systems.

The total number of electrons, determine the number of M—M bonds present, and predict a structure, indicated:

(a) μ—CO—[(η^4—C_4H_4)Fe(CO)]$_2$,

(b) μ—CO—μ—CRR′— [cp*Rh]$_2$,

(c) [μ—X—μ—CH_2—($Os_3(CO)_{10}$)]$^-$ where X = halide,

(d) (μ—Br)$_2$—[Mn(CO)$_4$]$_2$.

(*a*)

μ—CO	2 electrons neutral
2 (η^4—C_4H_4)	8 electronsneutral
2 CO	4 electrons neutral
2Fe(0)	16 electrons neutral
Total	30 electrons

M—M bonds = number of metals (n) × 2 – total no. of electron in complex/2 = [18(2) – 30]/2 = 3 Fe—Fe bonds

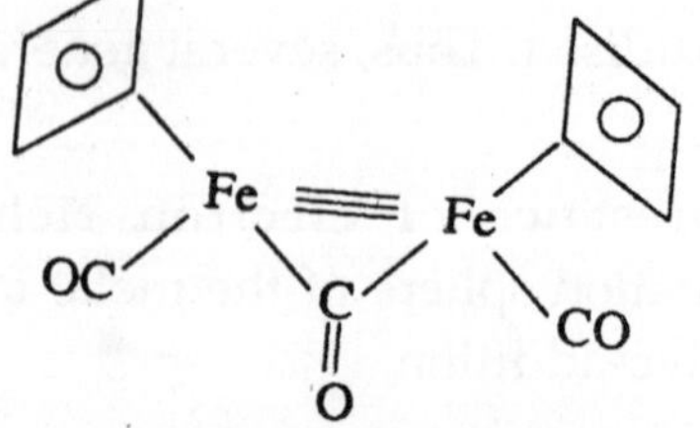

(*b*) μ—*CO* 2 electrons neutral

μ—CRR′ 2 electrons neutral

2[*cp**]$^-$ 12 electrons −2

2 Rh (I) 16 electrons +2

Total 32 electrons

M—M bonds = [18(2) − 32]/2 = 2 Rh—Rh bonds

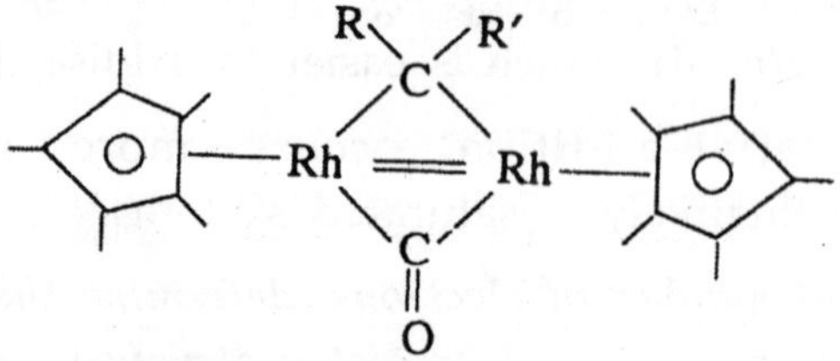

(*c*) [μ—X]$^-$ 4 electrons − 1

μ—CH_2 2 electrons neutral

10 CO 20 electrons

3 Os (0) 24 electrons neutral

Total 50 electrons

M—M bonds = [18(3) - 50]/2 = 2 Os—Os bonds

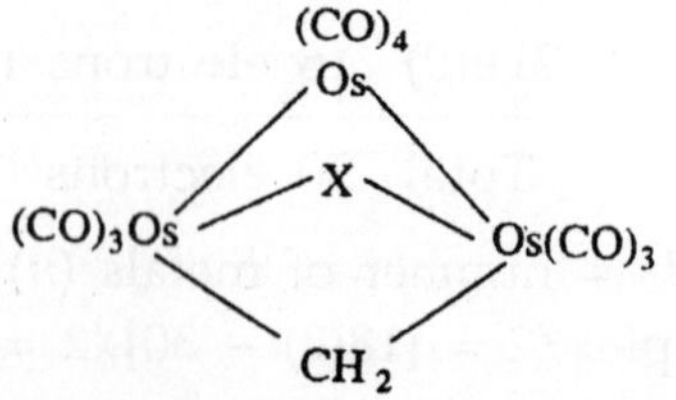

(*d*) 8 CO 16 electrons neutral

2 μ—Br 8 electrons – 2

2 Mn (I)— 12 electrons +2

Total 36 electrons

M—M bonds = [2(18) – 36]/2 = 0 Mn—Mn bonds

$(CO)_4Mn(\mu\text{-}Br)_2Mn(CO)_4$

For each of the following metal and ligand combinations, the simplest neutral compound that conforms to the 18-electron rule. Draw a reasonable structure, formulated:

(a) Fe, CO, COT = cyclooctatetraene; (b) Fe, cp, CO.

(*a*) $(CO)_3Fe(COT)$

3 CO 6 electrons neutral

Fe (0) 8 electrons neutral

COT 4 electrons neutral

Total 18 electrons

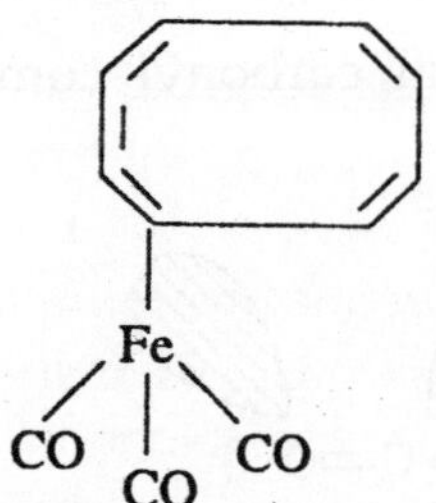

(*b*) $[cp\,(CO)_2Fe]_2$

4 CO 8 electrons neutral

2 cp^- 12 electrons -2

2 Fe(H) 12 electrons +4

Total 34 electrons plus one Fe—Fe bond

The structures of Isomers of Complex Dibromodicarbonyl Cyclopentadienyl-rhenium Drawn:

This is an example of geometrical isomerism in five coordinate complex, having square pyramid structure.

Factors Affecting the CO Stretching Frequency in the Infra-red Spectrum of Metal Carbonyl Derivatives:

The bonding in metal carbonyl complex is depicted below—

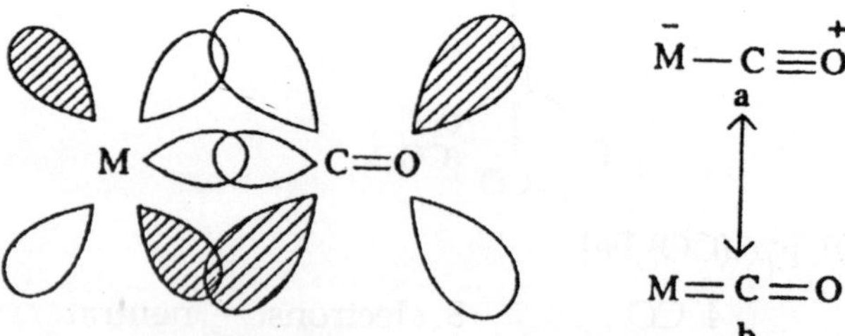

The pi bond between the metal and carbon atom arises as a result of a backbonding interaction between a filled metal rf-orbital of the appropriate symmetry and the empty pi star orbital of the carbonyl. Any factor that increases the electron density at the metal will increase the strength of the M—C pi bond, and thus the contribution from resonance form b. The result is a decrease in the CO bond order and a resulting decrease in the CO stretching frequency in the infra-red spectrum of the complex. Aspects which affect the DR. spectrum of metal carbonyl derivatives include the cf-electron count of the metal, the donor properties of the ligands and the charge on the complex.

For example, in case of $[W(CO)_5C1]^-$ and $Re(CO)_5Cl$, both metals are d^6 but since $[W(CO)_5C1]^-$ is anionic, it will have a lower CO stretching frequency. Similarly, in case of $Fe(CO)_5$ and $Fe(CO)_4Br_2$, $Fe(CO)_5$ will have a lower CO stretching frequency because being d^8 it has more electron density at the metal centre than the d^6 in another because of the lower oxidation state of the metal.

In the NMR spectrum of a metal complex at - 97° C, there are two signals, one at 182 ppm and 178 ppm. At 58° C the signals have just coalesced into a broad signal at 180 ppm. The rate constant for exchange of the two carbon atoms that produce these signals is:

At the coalescence temperature, the equation for the rate constant is

$$k = \frac{\pi(\Delta \nu)}{\sqrt{2}}$$

The difference in chemical shift of the two signals is 4 ppm, but the formula requires that this shift difference be in Hz. This value depends on the strength of the magnet used for the measurement and can be obtained by multiplying the operating frequency of the magnet (in MHz) and the difference in chemical shifts in ppm. Thus, for this example, we obtain a value of

$$\Delta\nu = 4 \text{ ppm } (25 \text{ MHz}) = 100 \text{ Hz}$$

and the value of the rate constant at – 58° C is calculated to be

$$k = \left(\pi / \sqrt{2}\right)\left(100\,s^{-1}\right)$$

$$k = 2.2 \times 10^2 s^{-1}$$

If a magnet of different field strength were used for this same measurement, the coalescence temperature would also be different.

Significant Reactions

The *mechanism of electrophilic substitution* reactions in ferrocene is illustrated below—

The attacking electrophile E^+ first interacts with Fe atom to give a π diene cationic intermediate which then loses a proton to give the final product, as shown—

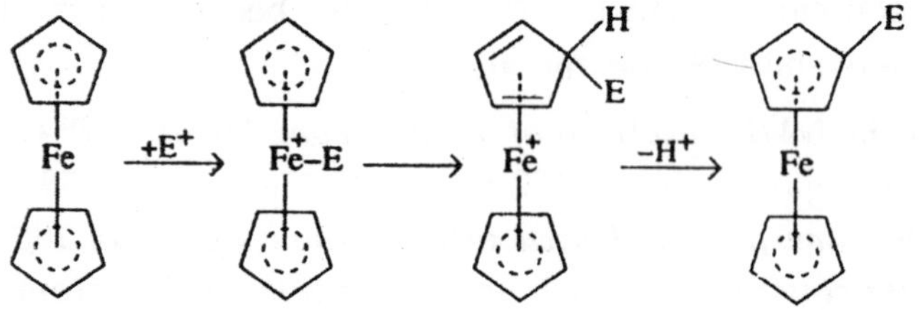

$$\text{Fe} \xrightarrow{+E^+} \overset{+}{\text{Fe}}\text{–E} \longrightarrow \overset{+}{\text{Fe}} \xrightarrow{-H^+} \text{Fe}$$

This mechanism explains why the rate of electrophilic substitution in ferrocene derivatives is proportional to the ease of oxidation of Fe atom.

Nitration and Halogenation

Direct nitration and halogenation leads to the decomposition of ferrocene. The processes are, therefore, carried out indirectly as described below—

Nitration:

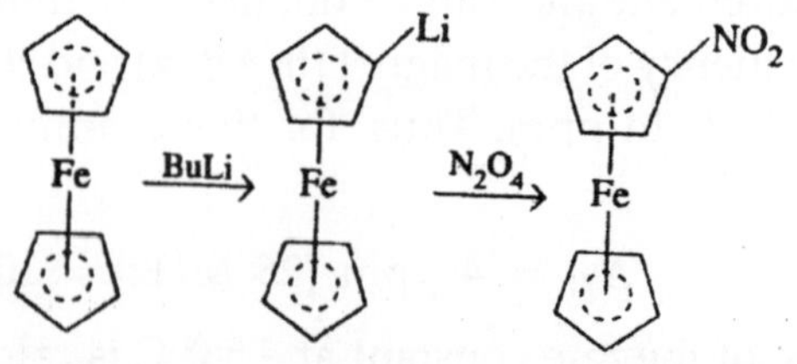

$$\text{Fe} \xrightarrow{\text{BuLi}} \text{Fe (Li)} \xrightarrow{N_2O_4} \text{Fe } (NO_2)$$

Bromination:

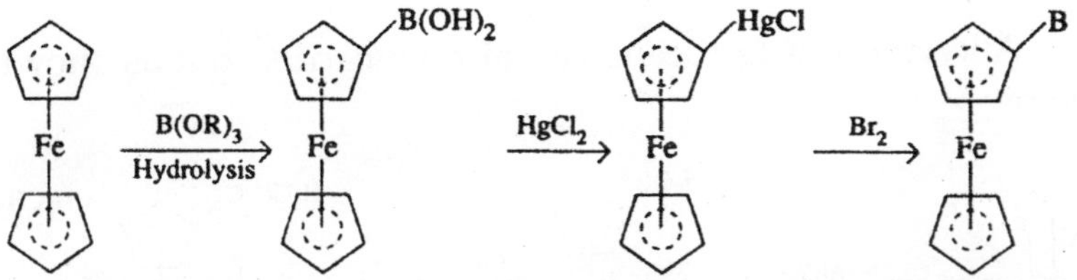

Nitroferrocene on reduction with Pd/HCl gives amino ferrocene.

Carboxylation: Ferrocene can also be carboxylated as shown—

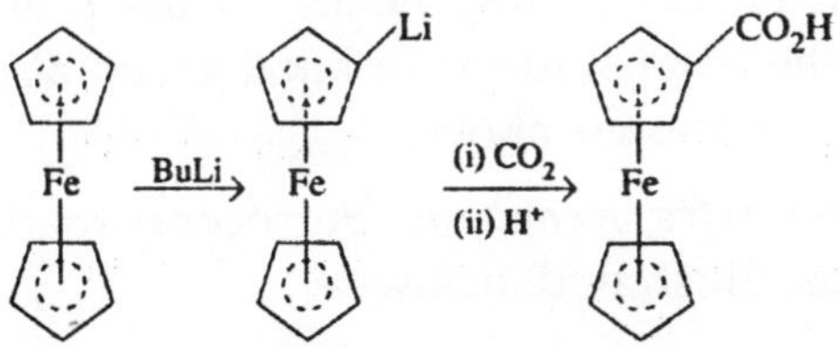

Vilsmeir Reaction: Ferrocene undergoes Vilsmeir reaction to yield ferrocene carboxyaldehyde which is a useful starting material for the preparation of other ferrocene derivatives.

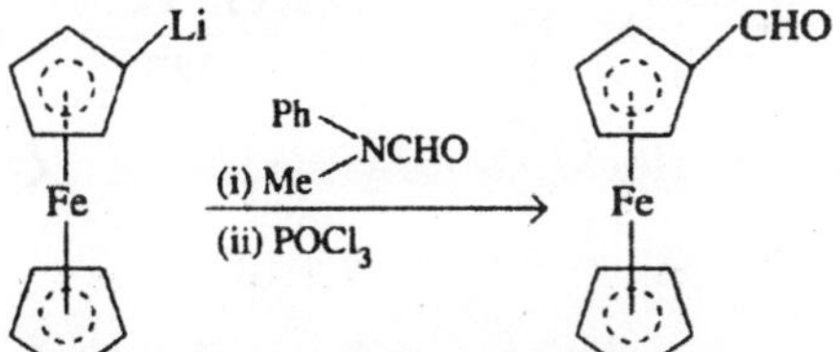

Mannich Condensation: Ferrocene also undergoes Mannich condensation as shown below—

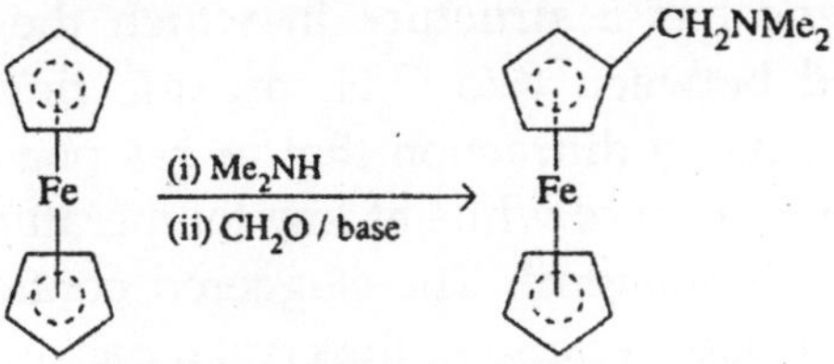

The Vilsmeir reaction and the Mannich condensation are reactions which are shown by aromatic systems that are more reactive than benzene. It follows, therefore, that ferrocene is more aromatic than benzene.

Alkyiation of ferrocene can also be carried out as shown below—

$$Fe(C_5H_5)_2 + H_2C=CH_2 \xrightarrow[230\ atm,\ 100°C]{AlCl_2} (H_3CH_2C\text{-}C_5H_4)Fe(C_5H_5) \xrightarrow[AlCl_2,\ 230\ atm\ 100°C]{Excess\ H_2C=CH_2} (H_3CH_2C)(CH_2CH_2)C_5H_3\text{-}Fe\text{-}C_5H_5$$

Since the presence of an alkyl group increases the electron density on the $C_5H_5^-$ ring, hence, in the presence of excess of ligand, the disubstitution of alkyl group also occurs in the same ring, as shown above.

Friedel-Crafts Reaction: Ferrocene readily undergoes acylation as illustrated below—

$$Fe(C_5H_5)_2 \xrightarrow[AlCl_3]{CH_3COCl} (C_5H_4COCH_3)Fe(C_5H_5) \xrightarrow[AlCl_3]{CH_3COCl} (C_5H_4COCH_3)Fe(C_5H_4COCH_3)$$

Ferrocene can also be alkylated in a similar manner.

Structure of Ferrocene

Ferrocene has a structure in which the iron atoms is sandwiched between two C_5H_5 organic rings. It has been observed by X-ray diffraction that in gas phase the structure of ferrocene is *eclipsed* while at low temperature the structure of ferrocene is *staggered.* The staggered configuration in the solid phase exists because of the crystal packing forces so that C—C and H—H repulsions between the two rings are minimum.

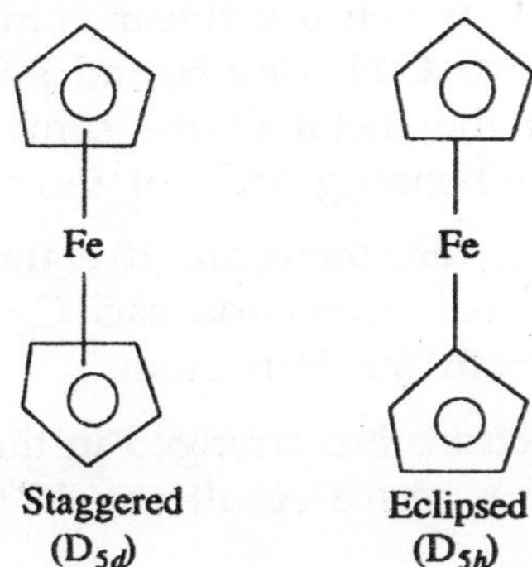

The five $p\pi$ AOs of the C_5H_5 ring combine linearly to give an equal number of MOs having symmetry a (i.e., one MO with no planar node), symmetry *eπ (i.e., two* degenerate MOs having one planar node) and symmetry e_2 *(i.e.,* two degenerate MOs having two planar node). The five MOs of C_5H_5 system are presented in Fig.

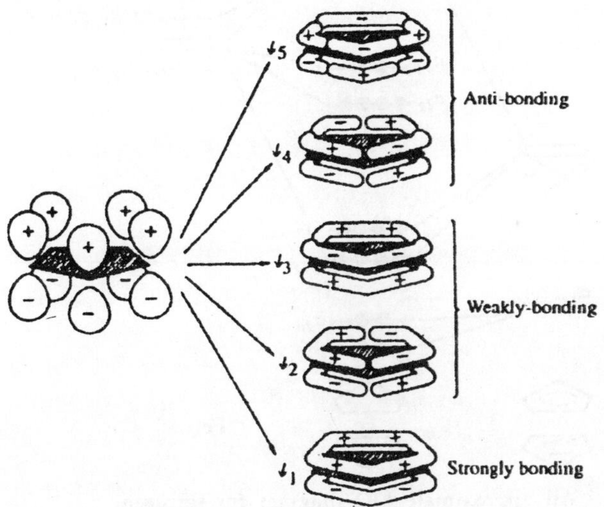

The *n* molecular orbitals formed from the set of *pn* orbitals of the C_5H_5 ring.

The MOs, *i.e.,* $\psi_1\psi_2$, etc. of one (*i.e.,* upper) C_5H_5 ring then linearly combine with MOs of the same symmetry, *i.e.,* $\psi'i$, ψ'_2, etc. of the lower C_5H_5 ring to given 10 MOs of resulting two C_5H_5 ring system.

The ten MOs which are linear combinations of ligand orbitals of the two C_5H_5 ring ligand system then combine with orbitals of the metal of the same symmetry to give bonding and antibonding MOs of ferrocene.

For $(\eta—C_5H_5)_2$ Fe, there are 18 valence electrons to be accommodated; 5π-electrons from each C_5H_5 ring and 8 valence shell electrons from the iron atom.

These 18 electrons are arranged in the nine bonding and non-bonding MOs. As a result the 10th antibonding MOs remain vacant.

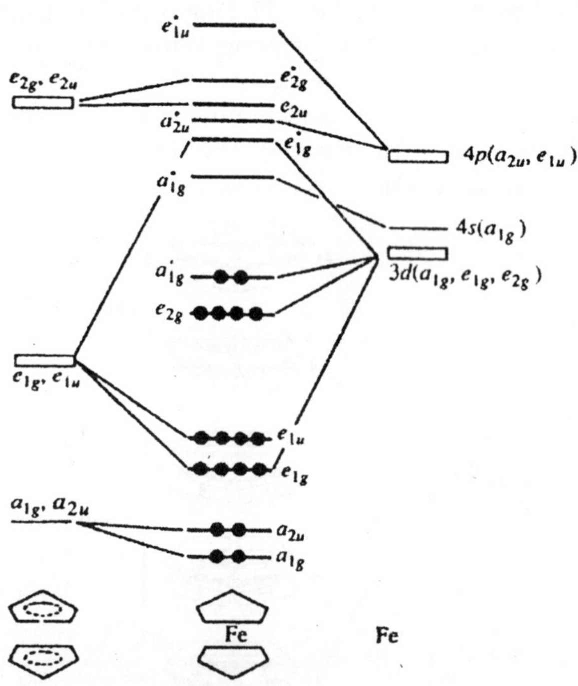

An approximate MO diagram for ferrocene

In the ferrocene molecule, the most significant bonds are the bonds of π-symmetry formed between Fe atom and the two C_5H_5 rings. These are obtained by the overlap of d_{xz} and d_{yz} AOs of Fe with appropriate ligand group MOs of C_5H_5 ring.

Ferrocene is diamagnetic in nature. Ferrocene has π-delocalised MOs and shows aromatic character. (If we consider the formation of ferrocene from Fe^{2+} and two $C_5H_5^-$ anions containing six *n*-electrons which obeys the 4*n*+ 2 rule of aromaticity and not the radical.)

$Fe(C_5H_5)_2$ is more stable than $Co(C_5H_5)_2$ or $Ni(C_5H_5)_2$. This is due to the stability of $Fe(C_5H_5)_2$ with 18 electrons compared to $Co(C_5H_5)_2$ with 19 electrons and $Ni(C_5H_5)_2$ with 20 electrons. Cobalocene and nikelocene have one and two electrons, respectively, in their ABMO whereas ferrocene has none.

The electrons in ABMO are easier to remove so that cobalocene and nickelocene are easily oxidisable compared to ferrocene.

Structure of Ruthenocene

For ferrocene the symmetry of the space group requires the two five membered rings to have staggered conformation. In ruthenocene and osmocene the rings adopt the eclipsed conformation.

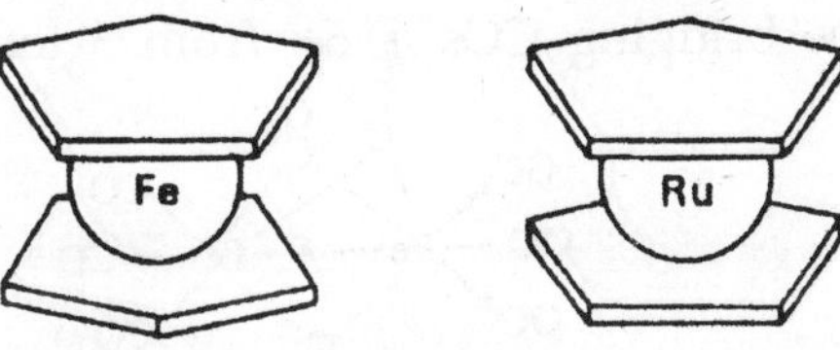

$[Co(H)(N_2)(PPh_3)_3]$:

N
≡
N

Ph_3P — Co — PPh_3

PPh_3

H

$[IrCl(CO)(O_2)(PPh_3)_2]$:

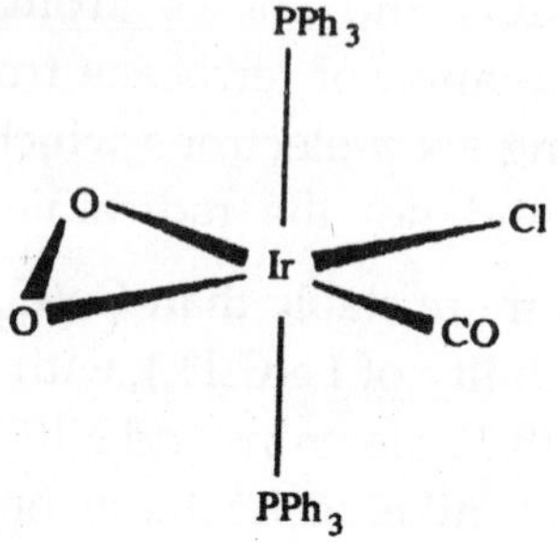

$[Ir^{III}Cl(CO)(O_2)(PPh_3)_2]$ complex

Structure of $Fe_2(CO)_9$**:** This carbonyl is formed by the photolysis of $Fe(CO)_5$. In the structure of this compound there is a weak coupling of the electron spins of Fe and Fe shown by dotted lines. This type of weak coupling of electron spins gives rise to a fractional single bond between two Fe atoms. These atoms are at a larger distance from each other than expected in a normal Fe—Fe single bond. The complex is diamagnetic and obeys the inert gas rule. Each Fe atom appears to have 36 electrons around itself ($26e^-$ from Fe + $3e^-$ from three bridging COs + $6e^-$ from 3 terminal COs.

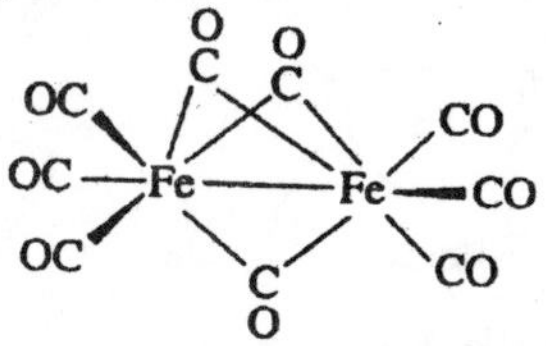

Structure of $Mn_2(CO)_{10}$: In the structure of this complex each metal atom has an octahedral environment and the two Mn(CO)5 units are joined solely by an M—M bond as shown. The equatorial CO groups of the two square planes are slightly bent towards each other. The Mn atom in half of the molecule has an effective atomic number of 35 (paramagnetic). One electron to become diamagnetic is provided by the M—M bond. Thus, diamagnetism of bi-metal (or bi-nuclear) carbonyls can be explained on the basis of the formation of an M—M bond in these carbonyls.

Structure of Zeise's Salt: It is a complex between platinum metal and ethylene, obtained when ethylene is passed through aqueous solution of potassium tetra chlroplatinate.

$$[PtCl_4]^{2-} + C_2H_4 \longrightarrow \underset{\text{Zeise's salt}}{[Pt(C_2H_4)Cl_3]^-} + Cl^-$$

In the structure of Zeise's salt, the ethylene occupies the fourth coordination site of the square planar complex with the C—C axis perpendicular to the platinum-ligand plane. In this compound, the dsp^2 hybridised σ orbital of Pt overlaps with π-bonding molecular orbitals of ethylene. Simultaneously, the filled $d\pi$ orbital of Pt overlaps with π^* orbital of C_2H_4.

The structure of the anion in Zeise's salt, trichloro (ethylene) platinate(II) ion.

Structure of $Re_2Cl_8^{2-}$: The complex ion octachloro-dirhenate(III) ion has short Re—Re distance and eclipsed configuration of the chlorine atom Re(III) is a d^4 species.

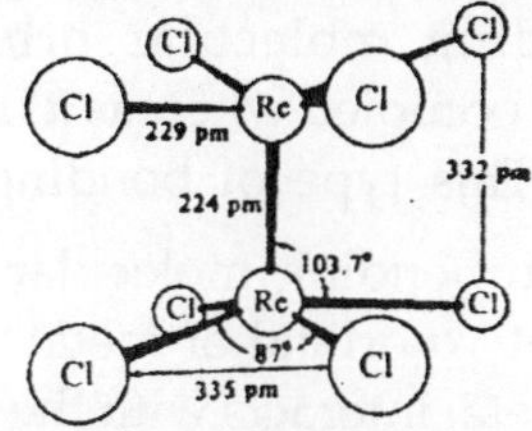

The structure of the octachlorodirhenate(III) ion, $Re_2Cl_8^{-2}$

The Re—Cl bonds may be considered tcrbe dative bonds from Cl^- ions to Re^{3+}. The formation of one sigma bond, two pi bonds and one delta bond causes the pairing of four electrons in quadruple bond, hence the complex is diamagnetic.

Structure of [Mo(C$_5$H$_5$)$_3$NO]: This molecule could be named *(monohaptocyclopentadienyl)* *(trihaptocyclopentadienyl)* *(Pentahapto focyclopentadienyl)* nitrosylmolybdenum or formulated as $(h^1—C_5H_5)(h^3—C_5H_5)(h^5—C_5H_5)$MoNO.

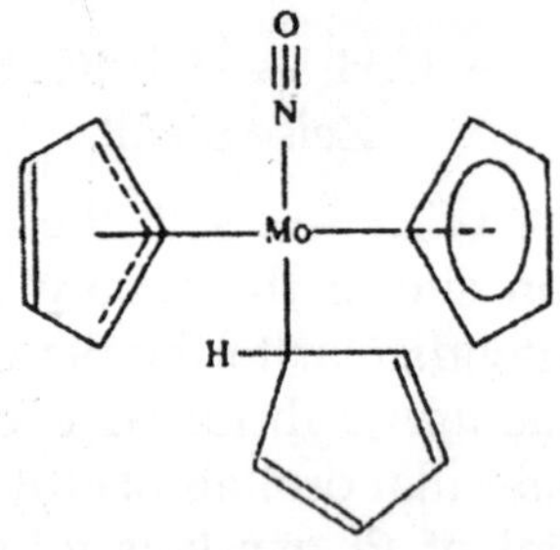

Proposed structure of $[Mo(C_5H_5)_3NO]$.

Explain the Nature of Metal-olefin Interaction: Organometallic complexes containing unsaturated ligands such as alkenes are considered olifinic complexes. Ziese characterised the pale yellow crystalline solid of composition $K[PtCl_3\ (C_2H_4)]$ which is now known as Zeise salt. This compound was prepared by passing ethylene gas through aqueous solution of potassium tetracholoro platinate.

$$PtCl_4^{2-} + C_2H_4 \longrightarrow [PtCl_3(C_2H_4)]^- + Cl^-$$

In the olifin ligand, in addition to a σ bond formed due to overlap of sp^2 hybrid orbitals of two carbon atoms, the p_z orbitals of the two carbon atom overlap to form a π-bonding and a π-antibonding molecular orbital. The π-bonding molecular orbital is occupied whereas π* antibonding molecular orbital is vacant. This type of bonding is shown in Fig.

The occupied rc bonding molecular orbital of olefin then overlaps with empty σ orbital of metal while one of the filled d. orbitals of the metal interacts with the empty π*-antibonding molecular orbital of (fin resulting in a π-symmetry about the Z-axis. Thus, the metal-olefin bond consists of two compounds, *viz.*, σ-donation from ligand to metal and simultaneous π-donation from metal to ligand.

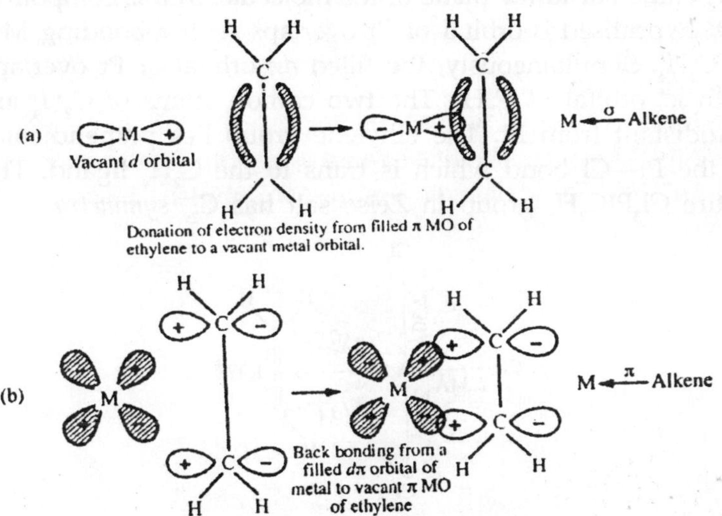

Stepwise bonding in metal-olefin complex.

Stepwise bonding in metal-olefin complex.

The metal carbon bond is, thus similar to metal-carbon bond in metal carbonyls. Both olefin and CO are weak σ-donors but the presence of low energy empty π* MO on these ligands makes the *d*π—π* bonding possible. The transfer of electrons from the ligand to metal during the formation of σ bond is enhanced by simultaneous removal of charge from the metal through *d*π—π* back donation. This effect is known as *'synergic effect'*, would strengthen the M—C bond.

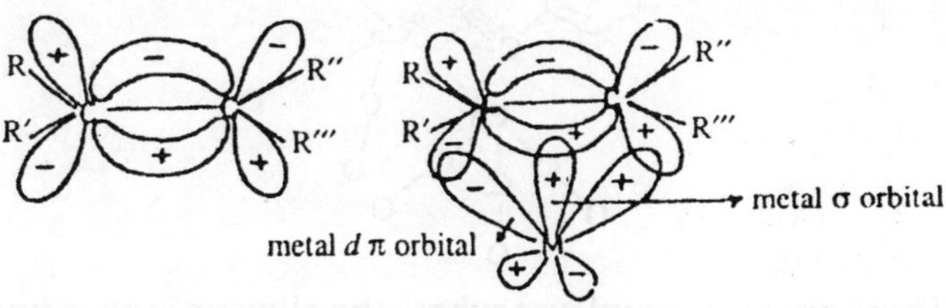

The structure of $K^+ [PtCl_3(C_2H_4)]^-$ has been established to approximately square planar. The olefin-metal bond lies perpendicular to the plane of the molecule. In this compound dsp^2 hybridised σ-orbital of Pt overlaps with π-bonding MO of C_2H_4. Simultaneously, the filled $d\pi$ orbital of Pt overlaps with π* orbital of C_2H_4. The two carbon atoms of C_2H_4 are equidistant from Pt. The ethylene group lies perpendicular to the Pt—Cl bond which is trans to the C_2H_4 ligand. The entire $Cl_3PtC_2H_4$ group in Zeise salt has C_{2v} *symmetry.*

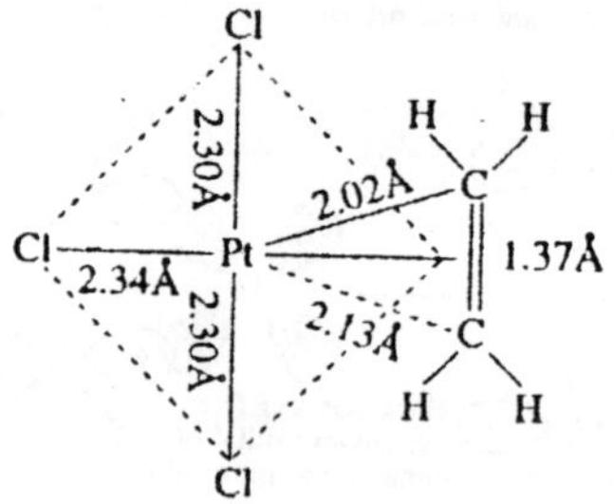

Alkyne Complexes

Alkynes can be bound to metals using one or both sets of π-bonds, as 2, 3 or $4e^-$ donors. Acetylene bridges of several types are known. In acetylene there are two π-bonds at 90° to each other and both can be bound to a metal. The Co atoms and the acetylene carbon atoms forms a distorted tetrahedron and the C_6H_5 or other groups $[Co_2(CO)_8(C_2R_2)]$ on the acetylene are bent away as shown

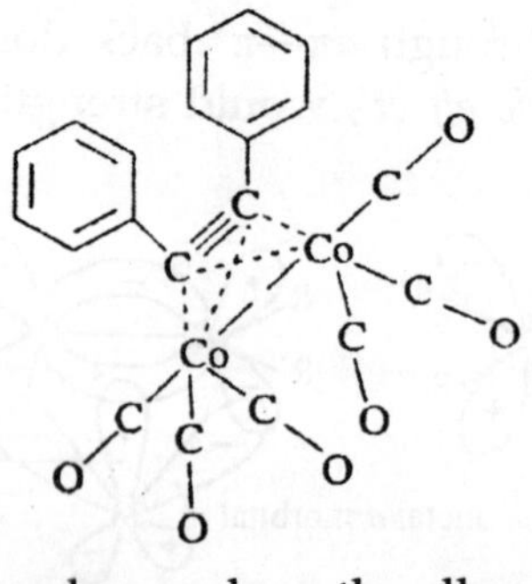

There are also complexes where the alkyene is coordinated to only one metal atom and serves simply as the equivalent

of the olefin or carbon monoxide ligand. Thus we have thereactions —

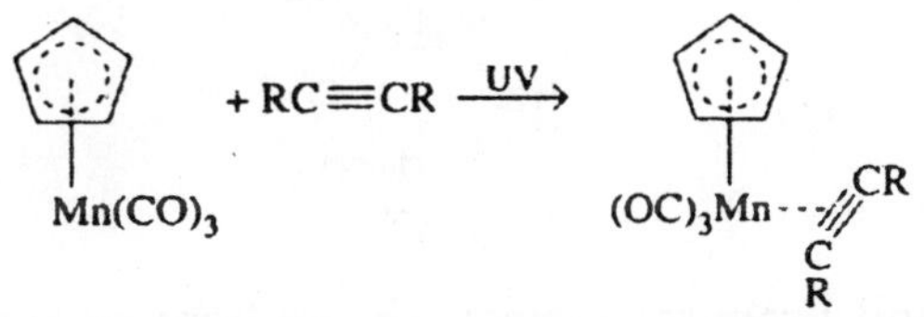

$(\eta^5\text{-}C_5H_5)Mn(CO)_3 + RC\equiv CR \xrightarrow{UV} (\eta^5\text{-}C_5H_5)(OC)_3Mn\cdots(RC\equiv CR)$ Where R = CF_3 or Ph

$PtCl_2^{2} + Bu^tC\equiv CBu^t \longrightarrow$ $[(Bu^tC\equiv CBu^t)ClPt(\mu\text{-}Cl)_2PtCl(Bu^tC\equiv CBu^t)]$

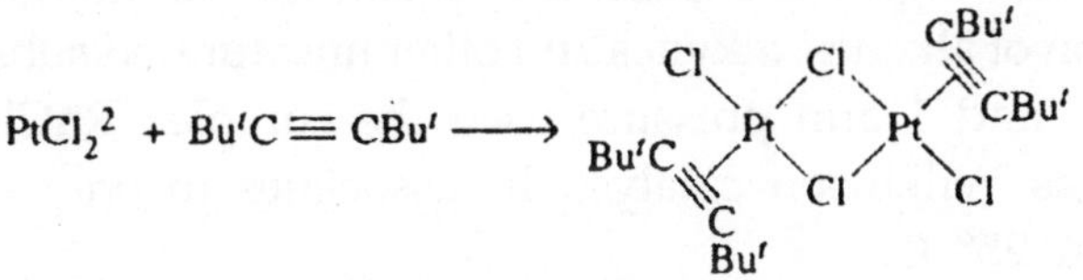

A third way of bonding, notably in Pt, Pd and Ir complexes is that shown below. In these the C—C stretching frequency is lowered considerably, to the range 1750 to 1770 cm ''' indicative of a C—C double bond.

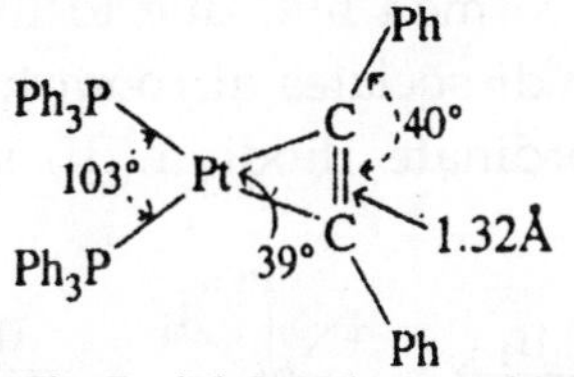

Ph_3P, Ph_3P, Pt, C, C, Ph, Ph, 103°, 39°, 40°, 1.32Å

The structure of $(Ph_3P)_2$ Pt (PhC_2PH), in which diphenylacetylene is most simply formulated as a divalent, bidentate ligand.

Many important reactions of acetylenes, especially with metal carboynyls, involve incorporation of the acetylenes into rings, thus producing species with new organic ligands bound to the metals. Some examples are the following—

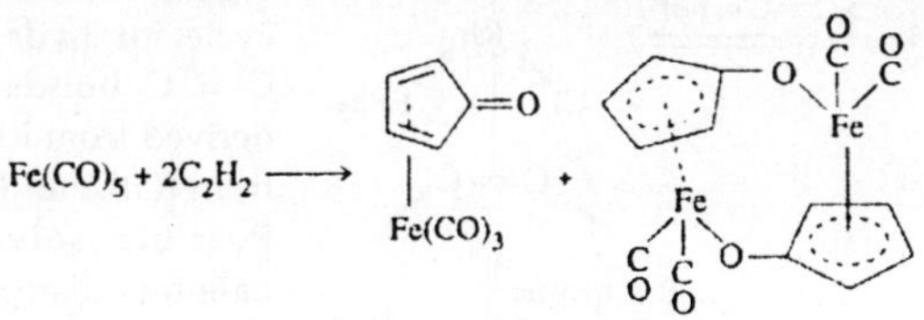

$Fe(CO)_5 + 2C_2H_2 \longrightarrow (C_4H_4CO)Fe(CO)_3 + [(C_5H_4)Fe(CO)_2\text{-}O]_2$-type dimer

$(\eta^5\text{-}C_5H_5)Co(CO)_2 + 2R_2C_2(R = CH_3 \text{ or } CF_3) \longrightarrow (\eta^5\text{-}C_5H_5)Co(C_4R_4CO)$

$$Fe(CO)_5 + 2C_2(CH_3)_2 \xrightarrow{h\nu}$$

Hydrogenation

The first rapid and practical system for the homogenous reduction of alkenes, alkynes and other unsaturated substances at 25° C and 1 atm pressure used the complex $RhCl(PPh_3)_3$ known as Wilkinson catalyst. It dissociates to only a small extent at 25° C.

$$RhCl(PPh_3)_3 \rightleftharpoons RHCl(PPh_3)_2 + PPh_3$$

Under hydrogen RhCl $(PPh_3)_3$ solution rapidly becomes yellow and ^1H and ^{31}P NMR studies show that an octahedral dihydride is first formed but, due to the strong trans effect of H, this rapidly dissociates at room temperature to give a fluxional five-coordinate rhodium(III) species.

This species then coordinates alkene.

Fig. Simplified catalytic cycle for hydrogenation of C = C bonds by species derived from $RhCl(PPh_3)_3$ or from $[(alkene)_2RhCl]_2 + PR_3$. Possible solvent coordination is disregarded. Cycle shows only major species involved for millimolar rhodium concentrations with large alkene concentrations under ambient conditions.

The 14-electron species $RhCl(PPh_3)_2$ adds a molecule of hydrogen oxidatively to form the 5-coordinate, 16-electron dihydrocomplex. This in turn adds a molecule of olefin to form a 6-coordinate, 18-electron complex.

Transfer of a hydrogen atom to the (β-carbon atom yields a 5-coordinate alkyl intermediate which then undergoes another rearrangement to give hydrogenated product and to reform the $RhCl(PPh_3)$ species.

The process is very slow because of the formation of stable rhodium ethylene complex, which does not readily undergo reaction with H_2.

$$(Ph_3P)_3RhCl + H_2C = CH_2 \longrightarrow (Ph_3P)_2RhCl(n^2—C_2H_4) + Ph_3P$$

Catalytic Cycle for the Production of Acetaidehyde

Wacker Process: The Wacker process is primarily used to produce acetaldehyde from the oxidation of ethylene by palladium(II)—Copper(II) chloride solution.

$$C_2H_4 + PdCl_2 + H_2O \longrightarrow CH_3CHO + Pd + 2HC1$$

$$Pd + 2CuCl_2 \longrightarrow PdCl_2 + CuCl$$

$$2CuCl + 2HC1 + \frac{1}{2} O_2 \longrightarrow 2CuCl_2 + H_2O$$

$$C_2H_3 + \frac{1}{2} O_2 \longrightarrow CH_3CHO$$

Oxidation of alkenes of the type RCH=CHR′ or $RCH=CH_2$ gives ketones, for example, acetone from propene.

The hydration of alkene—Pd(II) complex (a) occurs by the attack of H_2O from the solution on the concentrated ehtylene rather than the insertion of coordinated OH. Hydration, to form (b), is followed by two steps that isomerised the coordinated alcohol.

First (β-hydride eliminatiorf occurs with the formation of (c), and then migratory insertion results in the formation of (d). Elimination of the acetaldehyde and a hydrogen ion then leaves Pd (0). The Pd (0) is converted back to Pd(II) by the auxiliary Cu(II)-catalysed air oxidation.

The overall catalytic cycle is shown below—

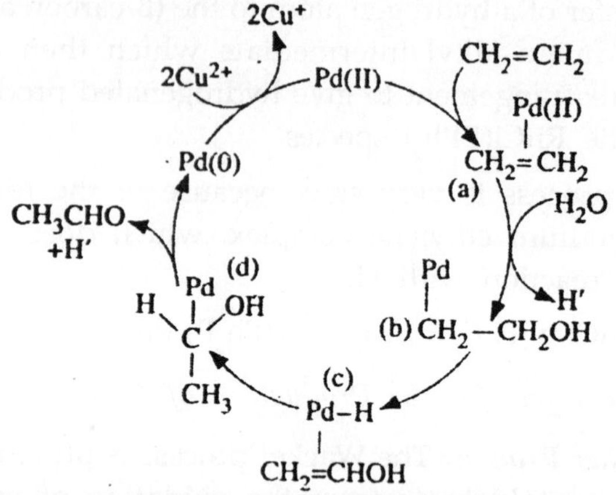

$2Cu^{+}$ | $2Cu^{2+}$ | Pd(II) | $CH_2=CH_2$ | Pd(II) | $CH_2=CH_2$ | (a) | H_2O | H' | Pd | (b) CH_2-CH_2OH | (c) | Pd–H | $CH_2=CHOH$ | CH_3 | H | C | OH | Pd | (d) | CH_3CHO $+H'$ | Pd(0)

Proposed mechanism for the oxidation of ethylene to acetaldehyde in the Wacker process. Chloride ligands have been omitted. The oxidation number of palladium is + 2 at all stages of this cycle except the upper left where reductive elimination of acetaldehyde gives Pd (0), which is oxidised by Cu (II). The complete cycle for the reoxidation of Cu (I) is not shown.

Acetone is prepared similarly from propene

$$CH_3CH=CH_2 + PdCl_2 + H_2O \xrightarrow{CuCl_2} CH_3-\overset{\overset{O}{\|}}{C}-CH_3 + Pd + 2\,HCl$$

Ziegler-Natta Polymerisation

The polymerisation of olefins by catalysts of the Ziegler-Natta type represents a most important example of the insertion reaction. The Ziegler catalyst formed from $TiCl_4$ and aluminium trialkyl. These component may react in a following way—

$$TiCl_4 + Al(C_2H_5)_3 \longrightarrow Al(C_2H_5)_2Cl + TiCl_3C_2H_5$$

$$TiCl_3C_2H_5 \longrightarrow HC1 + C_2H_4 + TiCl_2$$
$$TiCl_3C_2H_5 + HCl \longrightarrow C_2H_6 + TiCl_4$$
$$TiCl_2 + TiCl_4 \longrightarrow 2TiCl_4$$

In the inert hydrocarbon solvents used an insoluble mixed halide-alkyl complex of aluminium and titanium is formed of variable composition. This material is the active catalyst for the polymerisation of ethylene, persumably acting as a heterogenous catalyst.

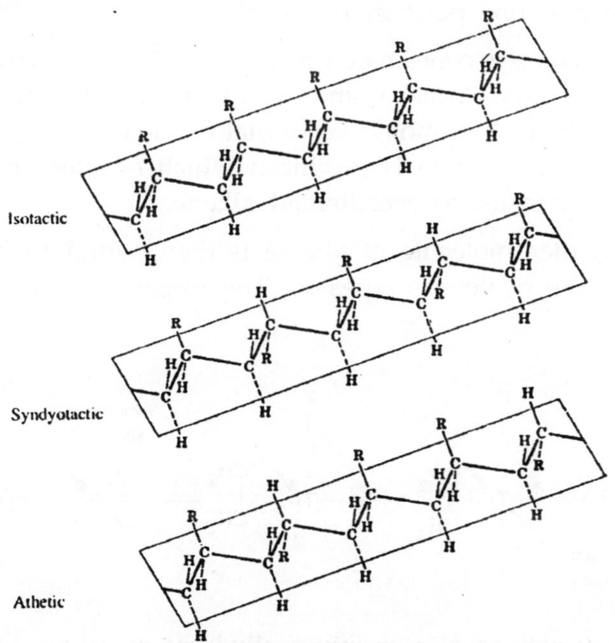

Different types of polymers

Natta and co-workers had produced stereospecific polymers. For example, olefins like propylene have been polymerised in such a way as to yield long linear head to tail chains consisting of sequences of monomer units having the same steric structures. These polymers are called *isotactic polymers* and they crystallise easily, whereas those monomeric units of different steric arrangement phased at random do not crystallise well. These polymers are called atactic. Polymers of regular, alternating structure are called *syndyotactic polymers.*

The Ziegler-Natta system is not homogenous, it is heterogenous and the active metal species is fibrous form of $TiCl_3$ formed in situ from $TiCl_4$ and $AlEt_3$ but performed $TiCl_3$ can be used. According to mechanism proposed by Cossee, the metal ion active site on the catalyst surface is octahedrally coordinated and which has a growing polymer chain R at one position and one position is vacant.

The alkene is coordinated to vacant site and an insertion reaction occurs. After promotion of an electron from the Ti—alkyl bond to a molecular orbital of the complex, a four centre transition state is produced which enables an alkyl group to transfer to coordinated alkene.

A further molecule of alkene is then bound to vacant site and the process is repeated. The mechanism is then as follows—

R
H_2C
Ti
C_2H_4
vacant site
R
H_2C
Ti
CH_2
CH_2
R
H_2C
Ti
CH_2
CH_2
vacant site
Ti
R
CH_2
CH_2
CH_2

The functions of aluminium alkyl are mainly—

(*i*) The formation of active metal species $TiCl_3$.

(*ii*) The replacement of one of the chloride ions in a Ti^{3+} ion at the surface by an alkyl radical derived from it.

The stereoregular polymerisation of propene may arise because of the nature of the sterically hindered surface sites on the $TiCl_3$ lattice.

An important extension of Ziegler-Natta polymerisation is the copolymerisation of styrene, butadiene and a third component such as dicyclopentadiene or 1, 4-hexadiene to give synthetic rubbers. Vanadyl halides rather than titanium halides are then used as the metal catalyst.

Wilke and his coworkers have shown that zera-valent complexes, especially of nickel, obtained by reduction with aluminium alkyls can be used in a wide variety of polymerisations such as trimerisation of butadiene to *trans,tran, trans-cyclododecatriene.*

Hydroformylation

Hydroformylation of alkenes (the 'oxo' reaction) is their conversion into aldehydes by the action of hydrogen and carbon monoxide at 90°-200° C temperature and 100-400 atm pressure in the presence of $Co_2(CO)g$.

$$>C=C< + CO + H_2 \xrightarrow{Co_2(CO)_8} H-\,>C=C<\,-CHO$$

Under these conditions $Co_2(CO)_8$ is converted into $HCo(CO)_4$ and strong evidence for this as the important cobalt containing species in the overall reaction is provided by its ability to bring about hydroformylation at ordinary temperature and pressure. The main steps in the mechanism are as follows—

$$HCo(CO)_4 \rightleftharpoons HCo(CO)_3 + CO \quad \text{...(i)}$$

$$RCH = CH_2 + HCo(CO)_3 \rightleftharpoons RCH_2CH_2Co(CO)_3 \quad \text{...(ii)}$$

$$RCH_2CH_2Co(CO)_3 + CO \rightleftharpoons RCH_2CH_2COCo(CO)_3 \quad \text{...(iii)}$$

$$RCH_2CH_2COCo(CO)_3 + H_2 \rightleftharpoons RCH_2CH_2CHO + HCo(CO)_3 \quad \text{...(iv)}$$

The eq. (ii) involves the β-hydrogen transfer to coordinated olefin, eq. (iii) the insertion of CO to form acyl and eq. (iv) the acyl cleavage by hydrogen. This process is frequently followed by reduction of the aldehyde to an alcohol.

The cobalt system is difficult to study. *The complex $RhH(CO)(PPh_3)_3$ is catalytically active even at 25° C and 1 atm pressure* and in contrast to cobalt system, produces only *aldehyde*. The catalyst cycle is shown in Fig. The initial step is attack of the alkene on the 16-electron species $RhH(CO)PPh_3$ (A) which leads to an alkyl complex (B) which then undergoes CO addition and insertion to form acyl derivative (C), which undergoes oxidative addition of molecular hydrogen to give the dihydrodoacyl complex (D). The high concentration of PPh_3 that are essential to provide high yield of linear aldehyde are probably required in order to suppress dissociation and the formation of monophosphine species and to force the associative attack of olefin on species (A).

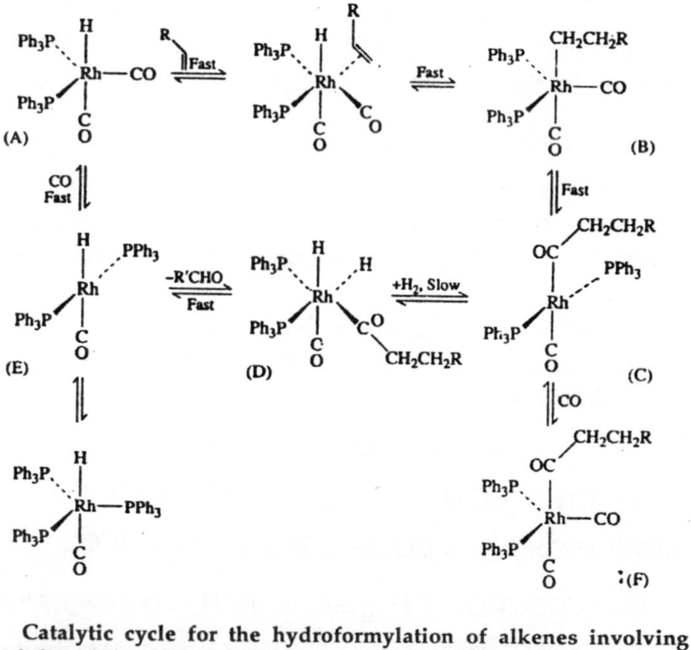

Catalytic cycle for the hydroformylation of alkenes involving triphenylphosphine rhodium complex species. Note that the configurations of the complexes are not known with certainty.

A wide variety of unsaturated substance can be hydroformylated by cobalt or rhodium catalysts but conjugated alkenes (*e.g.*, butadiene) may give a number of products including hydrogenated monoaldehydes. The mechanism is different, since addition of M—H to. dienes leads to alicyclic species which may be present as a-bonded intermediate or as η^3—allyls.

$$\begin{array}{c} CH_3 \\ / \\ M—CH \\ \backslash \\ CH \\ /\!/ \\ CH_2 \end{array} \rightleftharpoons \begin{array}{c} H—C—CH_3 \\ \vdots\;\backslash \\ M—\;C \\ \vdots\;/ \\ C \\ /\;\backslash \\ H\quad H \end{array}$$

Hydroformylation can also be achieved complexes such as $Ru(CO)_3(PPh_3)_2$ and $PtH(CO)(SnCl_3)$ $(PPh_3)_2$.

Isomerisation and Racemisation of Tris-chelate Complex

Tris-chelate complexes exist in enantiomeric configuration ^ and Δ about the metal atom, and when the chelating ligand is unsymmetrical, there are also geometrical isomers, cis and trans. Each geometrical isomer exists in enantiomeric forms; thus there are four different molecules.

Diagram of tris-chelate octahedral complexes (actual symmetry : showing how the absolute configuration Λ and Δ are defined according to the transla (twist) of the helices.

In the case of tris complexes with symmetrical ligands, the process of inversion (interconversion of enantiomers) is important. When the metal ions are of the inert type, it is often possible to resolve the complex; then the process of racemisation can be followed by measurement of optical rotation as a function of time. Possible pathways for racemisation fall into two broad classes; those without bond rupture and those with bond rupture.

There are two pathways without bond rupture. One is the trigonal or Bailor twist and other is rhombic, or Ray-Dutt, twist shown in Fig. Some cis-$M(CO)_4(PR_3)_2$ complexes are believed to isomerise in this way.

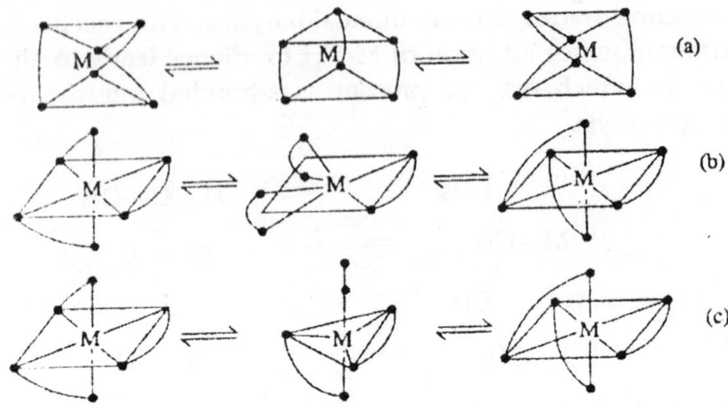

Three possible modes of intramolecular racemisation of a tris-chelate complex, (a) trigonal twist, (b) rhombic twist, (c) one of several ring-opening paths.

There are several possible dissociative pathway, in which a five-coordinated intermediate of either *tbp* or *sp* geometry are formed. There are no evidence for the actual occurence of associative pathway through seven coordinate intermediate.

It was found that both the isomerisation and the racemisation are intramolecular processes, which occur at approximately the same rate and with activation energies that are identical within experimental error. It thus appears likely that the two processes have the same transition state. This exclude a twist mechanism as the principal pathway for racemisation. Moreover, it was found that isomerisation occurs mainly with inversion of configuration.

This imposes a considerable restriction on the acceptable pathways. Detailed consideration of the stereochemical consequences of the various dissociative pathways, and combinations thereof, leads to the conclusion that for this system the major pathway is through a *tbp* intermediate with the dangling ligand in an axial position.

Isomerisation

There are many instances of the isomerisation of alkenes or alkenilic compounds under the influence of transition metal compound catalyst. Some examples are—pent-1-ene into cis- and trans-pent-2-ene ($RhCl_3$ in methanal or H_2PtCl_6 + $SnCl_2$); hex-1-ene into hex-2-enes and hex-3-enes ($Fe_3(CO)_{12}$); penta-l,4-diene into trans-penta-l,3-diene$Fe(CO)_3Fe(CO)_5$; allyl alcohol into propionaldehyde ($Fe(CO)_5$ or $HCo(CO)_4$) propylene oxide into acetone ($Co_2(CO)_8$ in methanal) and isopropyl magnesium bromide into «-propyl compound ($TiCl_4$).

In these processes it is believd that the double bond reacts with transition metal complex, giving an intermediate which decomposes with the regeneration of the double bond in a new position. Possible mechanism of such changes are shown below—

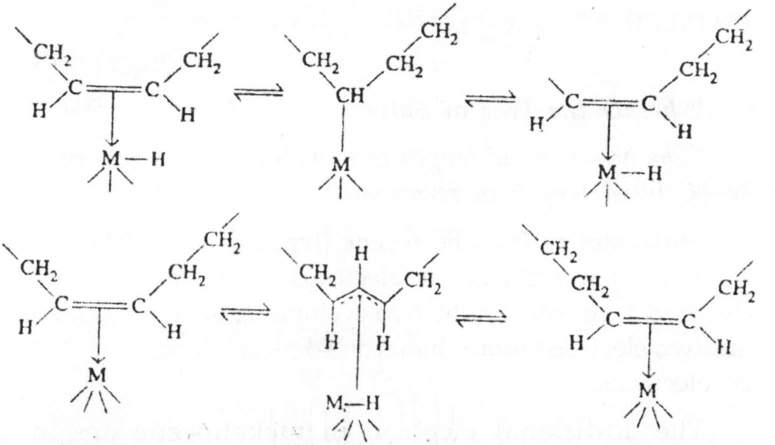

Isomerisation normally leads to the isomer or mixture of isomers predicted by thermodynamic considerations. For the isomerisation of allyl alcohol to propionaldehyde by $HCo(CO)_4$ the following mechanism has been suggested—

$$CH_2 = CHCH_2OH + HCo(CO)_4 \longrightarrow \begin{array}{c} CH_2 = CHCH_2OH + CO \\ \downarrow \\ HCo(CO)_3 \end{array}$$

$$\longrightarrow CH_3-\overset{H}{\underset{Co(CO)_3}{C}}-\overset{H}{\underset{H}{C}}-OH \longrightarrow CH_3CH=CHOH + HCo(CO)_3 \rightarrow CH_3CH_2CHO$$

(which reacts with more allyl alcohol)

Evidence in support of this mechanism is provided by using the deuteride $DCo(CO)_4$ as catalyst, when CH_2DCH_2CHO is the only deuterium-containing organic product of the reaction. In the isomerisation of isopropylmagnesium bromide the mechanism appears to be—

$$(CH_3)_2CHMgBr \xrightarrow{TiCl_4} (CH_3)_2CHTiCl_3 \longrightarrow TiCl_3H + CH_3CH = CH_2$$

$$CH_3CH = CH_2 + TiCl_3H \rightleftharpoons CH_3CH_2CH_2TiCl_3$$

$$CH_3CH_2CH_2TiCl_3 + (CH_3)_2CHMgBr \rightleftharpoons (CH_3)_2CHTiCl_3 + CH_3CH_2CH_2MgBr$$

Whether the True or False

"The Ni—C bond length in nickelocene is longer than the Fe—C bond length in ferrocene."

Statement is True: Ferrocene $[(cp)_2Fe]$ has 18 electrons in the molecular orbitals: six electrons are from Fe^{2+} and six electrons from each of the two cyclopentadiene anions. Nickel has two electrons more than iron so nickellocene $[(cp)_2Ni]$ has 20 electrons.

The additional electron in nickellocene are in an antibonding orbital. The bond order of nickellocene is higher than that of ferrocene hence Ni—C bond length in nickellocene is larger (2.20 A) than the Fe—C bond length in ferrocene (2.06 A).

Structure of $Co_4(CO)_{12}$: One metal atom has three terminal CO groups, but the remaining nine CO groups occupy both symmetrical bridging positions and terminal positions around the triangle formed by the other three metal atoms. The molecule has C_{3v} symmetry.

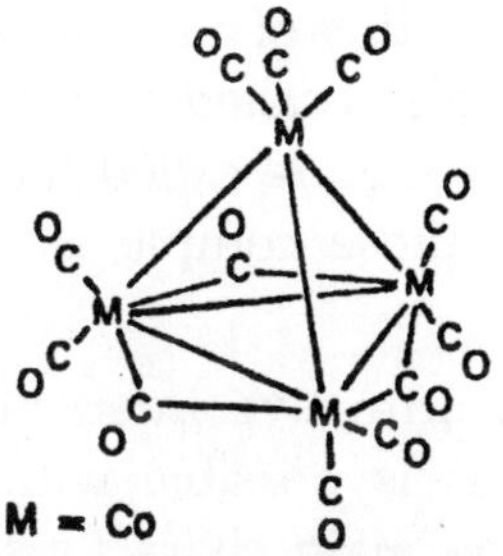

M = Co

Isolobal Relationship: Two molecular fragments are isolobal if the number, symmetry properties, shapes and approximate energies of their frontier orbitals are the same. They may or may not also be isoelectronic. If the two fragments are isoelectronic, in the sense that the ratio of the number of electrons in frontier orbitals to the number of frontier orbitals is same, then the net charges on the two species will be the same. The orbitals whose similarity is critical in determining isolobality are called frontier orbitals.

Cobalt fragment $Co(CO)_3$ is isolobal with —CH.

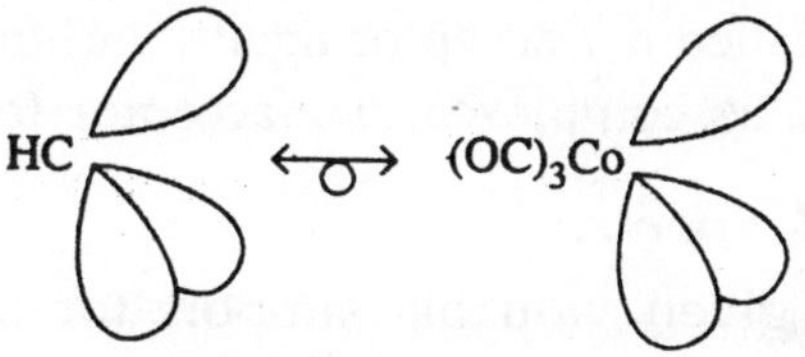

HC ⟷ $(OC)_3Co$

The cluster formed after replacing $Co(CO)_3$ fragment by isolobal CH is $Co_3(CO)_9CH$, with the structure :

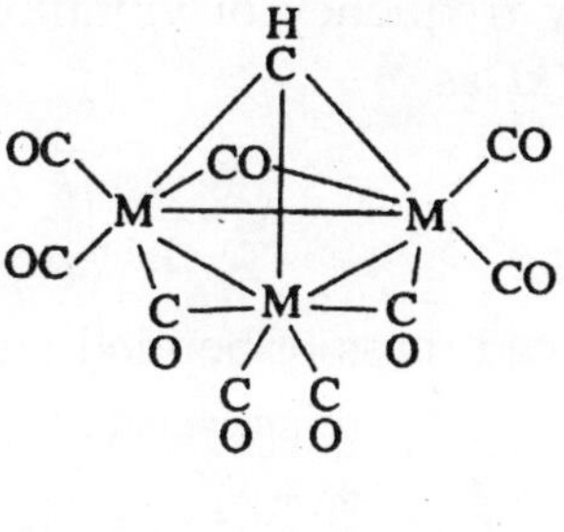

***n*-acidity:** The ability of a ligand to accept electron density into vacant n orbitals is called n-acidity. Complexes of metals with carbon monoxide are called metal carbonyls. Oxidation state of metal in these complexes is either zero or a low positive value.

There are no attractive interactions between the metal and the ligands as is possible with the positively charged metal ion. It is the main characteristic of CO ligand that it can stabilise low oxidation states. This is due to the fact that it possesses vacant π orbitals in addition to lone pairs. The formation of a sigma bond by the donation of a lone pair of electrons into the suitable vacant metal orbitals leads to excessive negative charge on the metal (in zero or negative oxidation state). To counter the accumulation of negative charge on the metal, a π-bond is formed by the back donation of electrons from the filled metal orbitals into the vacant π-type orbitals on the ligand. This also supplements the σ-bond. This ability of the ligand (CO) to accept electron density into vacant π orbitals is called π-acidity. Therefore, CO is also called a π-acceptor ligand and the metal carbonyls are referred as complexes of π-acceptor (or π-acid) ligands.

Vibrational Spectra

IR has given valuable support for bonding in metal carbonyls. These studies provide information regarding bond orders of M—C and C≡O bonds. The decrease in C—O bond order or force constant is estimated by studying the CO stretching frequency in IR spectroscopy. The IR spectra is characterised by frequency of vibration, which is related to force constant *(k)* as

$$V = \frac{1}{2\pi}\sqrt{\frac{k}{\mu}}$$

where μ is reduced mass of bonded atoms with mass m_1 and m_2 and is given by $\mu = \frac{m_1 m_2}{m_1 + m_2}$

The force constant is a measure of bond strength. The larger the force constant, the stronger the bond and higher the vibrational frequency. In IR spectrum CO frequencies are generally very strong and, therefore, can be used to study the force constant or bond strength of CO bond in different bonding situation.

Free CO has IR stretching frequency of 2143 cm^{-1}. In the case of terminal CO groups in neutral molecules, the CO frequency has been found to be 2125-1850 cm^{-1}. This suggests that the bond strength of CO has decreased or M—C bond strength has increased. This is only possible if there is back bonding from filled metal *d*-orbitals to the antibonding orbitals of CO.

Possible Isomers of $cp_2Fe_2(CO)_4$

Geometrical Isomers of cp_2Fe_2 (CO)4:

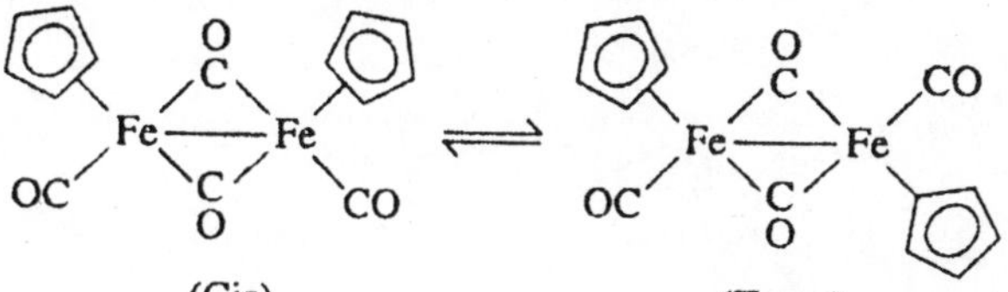

(Cis) ⇌ (Trans)

The total number of electrons per Mo atom $= \frac{34}{2} = 17 = 17$.

Thus the effective atomic number of each Mo atom is one short of 18. So there is an Mo—Mo bond and the bond order is one.

Assuming the coordination number of Mo as 6, $Mo_2(cp)_2(CO)_6$ has bridge structure, the cis and trans isomers can be drawn as—

(Cis) (Trans)

Elements not in Transition

The Structural Features of: **(a) B_4H_{10} (b) B_5H_9 (c) B_5H_{11}**

(a) *Tetraborane, B_4H_{10}:* In this molecule four B-atoms may be regarded as a portion of slightly distorted octahedron. The structure of this molecule contains:

(i) Fourt bridging (3*c*—2*e*) B—H—B bonds

(ii) One direct (2*c*—2*e*) B—B bond

(iii) Six terminal (2*c*—2*a*) B—H bonds

The structure can be represented as follows:

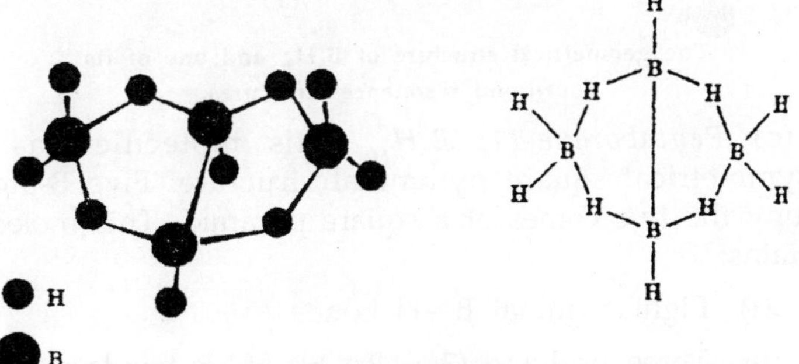

The structure and line representation of the bonding in tetraborane, B_4H_{10}.

Thus six B—H electron pair, one B—B electron pair and four B—H—B bridge bonds together account for twenty-two valence electrons contributed by four boron and ten hydrogen atoms. Thus skeletal electrons are 22.

(b) *Pentaborane-9, B_5H_9*: In this moleculé five B-atoms form a square-based pyramid. Four B-atoms located in the base of the square-pyramid are bonded to each other by four (3c – 2e) B—H—B bonds while the B-atom located at the apex of the pyramid is bonded to the two basal B-atoms (2c – 2e) B—B bond. This molecule also contains one closed (3c - 2e) B—B—B bond. Thus B_5 H_9 molecule contains:

(i) Five terminal B—H bonds

(ii) Four bridging B—H—B bonds

(iii) Two B—B bonds

(iv) One closed B—B—B bond

Total 24 skeletal electrons are involved in various bonds formation in the molecule.

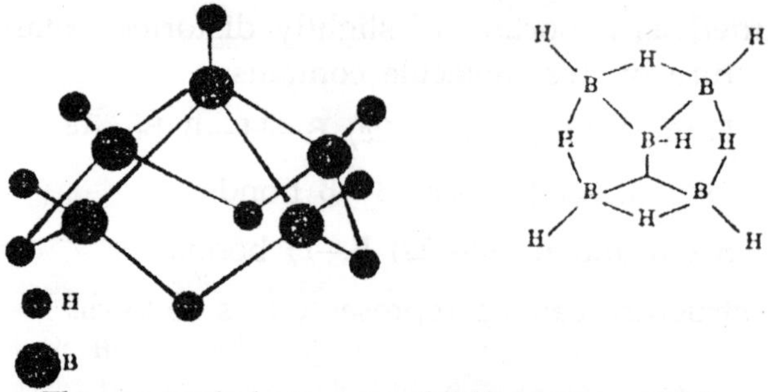

The geometrical structure of B_5H_9, and one of its four-bond resonance structures.

(c) *Pentaborane-11, B_5H_{11}:* This molecule has an unsymmetrical square pyramidal structure. Five B-atoms occupy the five comes of a square pyramid. This molecule contains:

(i) Eight terminal B—H bonds

(ii) Three bridging (3*c* - 2*e*) B—H—B bonds

(iii) Two closed (3*c* - 2*e*) B—B—B bonds

The structure is shown below:

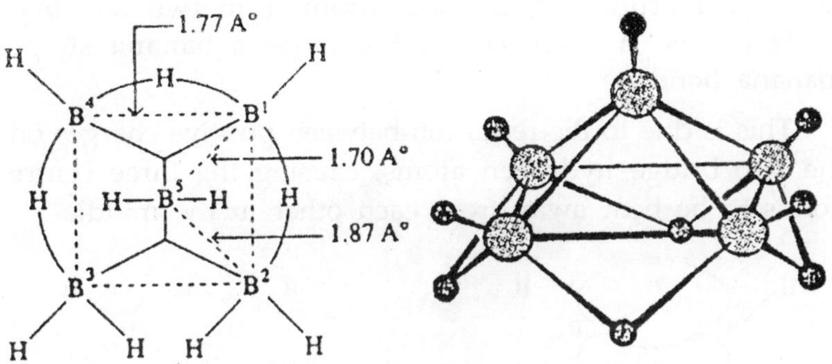

Structure of pentaborane-11, B_5H_{11}

Total number of skeletal electrons are 26.

The wave functions of molecular orbitals of diborane in terms of orbitals ϕ_{B_1}, ϕ_{B_2}, **and** ϕ_H.

The hydrogen bridge structure of *diborane* (B_2H_6) is shown in Fig.

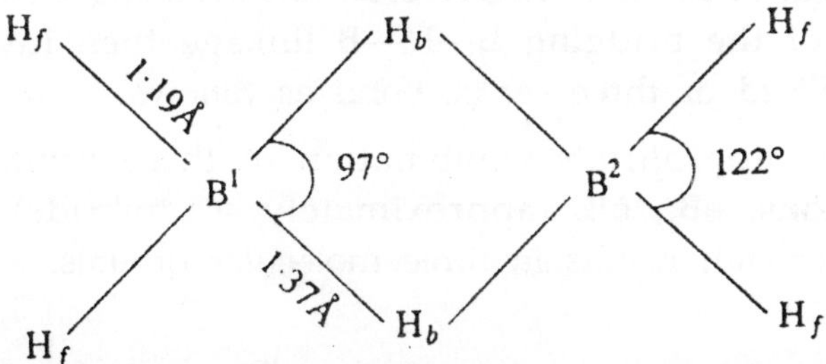

It contains two type of bonds:

(i) Four terminal (2c - 2e) B—H bonds (normal covalent a bonds)

(ii) Two bridging (3c - 2e) B—H—B bonds

In terms of molecular orbitals, the three centre B—H—B orbital may be considered to result from the combination of one sp^3 orbital from each borane and the j-orbital of the hydrogen. Thus two bridging (3c - 2e) B—H—B bonds are formed by two of the four sp^3 hybrid orbitals, one of them is empty and other is singly filled. Other two singly filled sp^3 hybrid orbitals of a boron atom form two terminal B—H bonds. A three centre bond has a banana shape (banana bond).

This is due to the repulsion between positive charges on the two bridge hydrogen atoms, causing the three centre bonds to be bent away from each other in the middle.

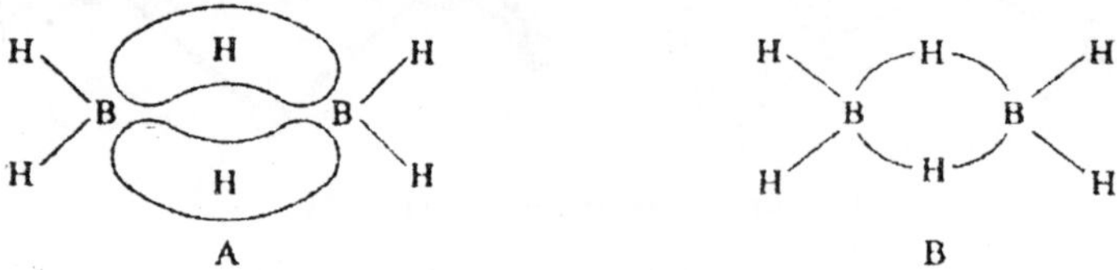

(A) Qualitative picture of bonding in diborane. (B) A common method of depicting B—H—B bridges

Wave Mechanical Picture of Diborane: Consider each borane atom to be sp^3 hybridised. Two terminal B—H bonds are cr bonds involving a pair of electron each. This accounts for eight of the total twelve electrons available for bonding. Each of the bridging B—H—B linkage then involves a delocalised or three centre bond as follows.

The appropriate combination of three orbital wave functions, ϕB_1, ϕB_2 (approximately sp^3 hybrids) and ϕ_H (an s-orbital) results in three molecular orbitals.

$$\text{Bonding } \psi_b = \frac{1}{2}\phi B_1 + \frac{1}{2}\phi B_2 + \frac{1}{\sqrt{2}}\phi_H$$

$$\text{Non-bonding } \psi_n = \frac{1}{\sqrt{2}}\phi B_1 - \frac{1}{\sqrt{2}}\phi B_2$$

$$\text{Antibonding } \psi_n = \frac{1}{2}\phi B_1 + \frac{1}{2}\phi B_2 - \frac{1}{\sqrt{2}}\phi_H$$

The diagrammatically possibilities of overlap together with the resulting MOs and their energies are given in Fig:

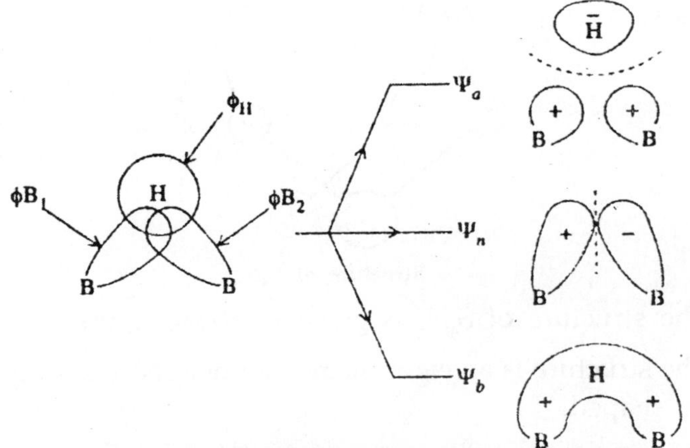

Qualitative description of atomic orbitals (left), resulting three-centre molecular orbitals (right), and the approximate energy level diagram (centre) for one B—H—B bridge in diborane.

The preparation, properties and structure of S_4N_4:

S_4N_4, Tetrasulphur Tetranitride: It is prepared as follows:

$$6SCl_2 + 16NH_3 \longrightarrow S_4N_4 + 2S + 14NH_4Cl$$

$$6S_2Cl_2 + 16NH_3 \xrightarrow{CCl_4} S_4N_4 + 8S + 12NH_4Cl$$

$$6S_2Cl_2 + 4NH_4Cl \longrightarrow S_4N_4 + 8S + 16HCl$$

The compound is also formed when sulphur reacts with anhydrous liquid ammonia.

$$10\ S + 4\ NH_3 \rightleftharpoons S_4N_4 + 6\ H_2S$$

S_4N_4 is solid, m.p. 178°C. It is *thermochromic,* that it is changes colour with temperature. At liquid nitrogen temperatures it is almost colourless, but at room temperature it is orange-yellow, and at 100°C it is red.

S_4N_4 is very slowly hydrolysed by water, but reacts rapidly with warm NaOH with the break-up of the ring:

$$S_4N_4 + 6NaOH + 3H_2O \longrightarrow Na_2S_2O_3 + 2Na_2SO_3 + 4NH_3$$

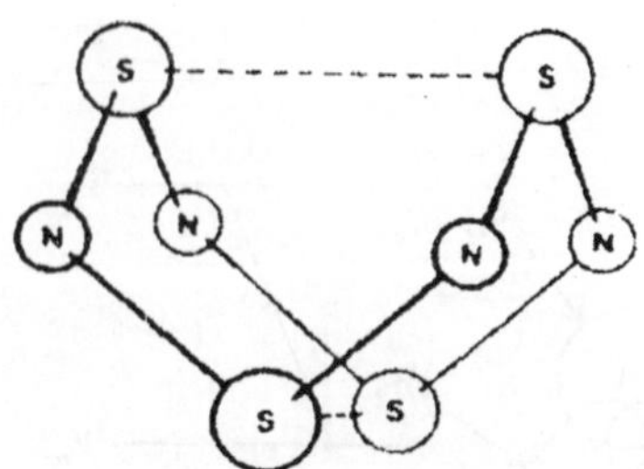

Structure of S_4N_4

The structure of S_4N_4 is given in above figure.

The structure is an eight membered heterocyclic ring and cradle shaped.

Since decomposition of tetrasulphur tetranitride with alkali gives ammonia, and reduction with stannous chloride followed by decomposition with alkali also gives ammonia, there can be no N—N links in molecule.

If the group N—N were present, direct decomposition would yield ammonia and reduction followed by decomposition would yield hydrazine or its derivatives.

S_4N_4 is diamagnetic S—N bond lengths are 1.74Å,

S—S distance = 2.63Å,

N—N distance = 1.47.

Band angles are SNN = 110°,

SNS = 98°, NSN = 76°.

$B_{10}C_2H_{12}$ is isostructural and isoelectronic with what borane ion $B_xH_y^{2-}$:

$B_{10}C_2H_{12}$ is isostructural and isoelectronic with $B_{12}H_{12}^{2-}$ borane ion. The boron atom each have one fewer electron than a carbon atom. To keep the total number of electrons the same in $B_{10}C_2H_{12}$ and the borane ion, the replacement of two carbon atoms with boron atoms must be accompanied by addition of two extra electrons.

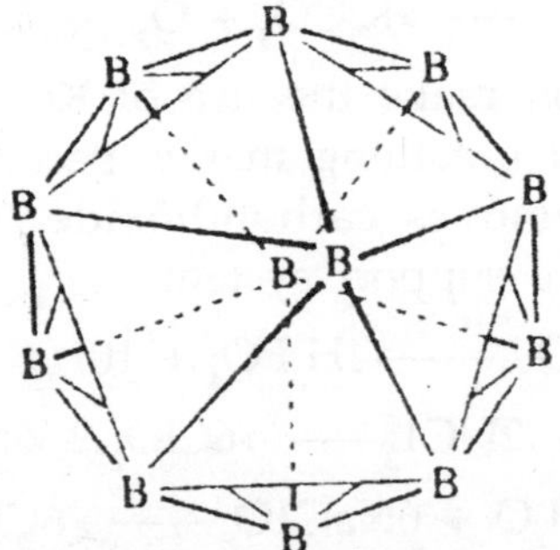

Framework of the$B_{12}H_{12}^{2-}$ ion

Reactions in Balance

(i) *Reaction of potassium superoxide with carbon dioxide*

(ii) *Hydrolysis of phosphorus sulphide*

(iii) *Reaction of sodium chlorite with nitrogen trichloride*

(iv) *Hydrolysis of calcium cyanamide*

(v) *Reaction of borontrioxide with cobalt oxide*

(vi) *Hydrolysis of phosgene*

(vii) *Reaction of carbon tetrachloride with hydrogen fluoride in anhydrous conditions.*

(viii) *Reaction of hydrazine with zinc in acidic medium*

(ix) *Diborane reacts with ammonia*

(x) *Silica reacts with carbon in an electric furnace*

(xi) *Reaction of ammonia with disulphur dichloride*

(xii) *Reaction of bromate and chloride in acid medium*

(xiii) *Hydrolysis of nitrogen trichloride*

(xiv) *Reaction of bromate and bromide in acid solution*

(xv) *Hydrolysis of borazine*

(xvi) *Hydrolysis of silicon tetrafluoride*

(xvii) *Hydrolysis of tetrasulphur tetranitride*

(i) $KO_2 + 2CO_2 \longrightarrow K_2CO_3 + O_2$

This reaction make the use of KO_2 in space capsules, submarines and breathing marks, because i both produces oxygen and removes carbondioxide. Both functions are important in life support system.

(ii) $P_4S_{10} + 16H_2O \longrightarrow 4H_3PO_4 + 10H_2S$

(iii) $6NaClO_2 + 2NC1_3 \longrightarrow 6ClO_2 + 6NaCl + N_2$

and$NCI3 + 3H_2O + 6NaC1O_2 \longrightarrow 6ClO_2 + 3NaCl$

$+ 3NaOH + NH_3$

(iv) $CaNCN + 3H_2O \longrightarrow 2NH_3 + CaCO_3$

(v) $CoO + B_2O_3 \longrightarrow Co(BO_2)_2$

Cobalt metaborate

Blue colour

(vi) $COC1_2 + H_2O \longrightarrow 2HC1 + CO_2$

Phosgene

(vii) $CCl_4 + 2HF \xrightarrow[\text{SbCl}_5\text{ Catalyst}]{\text{Anhyd condition}} CCl_2 + F_2 + 2HCl$

Feron

(viii) $N_2H_4 + Zn + 2HCl \quad 2NH_3 + ZnCl_2$

Hydrazine (-II)

(ix) $\underset{\text{Diborane}}{B_2H_6} + 2NH_3 \xrightarrow[\text{low temp.}]{\text{Excess NH}_3} B_2H_6.2NH_3$

$\xrightarrow[\text{high temp.}]{\text{Excess NH}_3} \underset{\text{Boron nitride}}{(BN)_x}$

$\xrightarrow[\text{high temp.}]{\text{Ratio 2NH}_3\text{:1B}_2\text{H}_6} \underset{\text{Borazine}}{B_3N_3H_6}$

(x) $SiO_2 + 3C \longrightarrow Si + 2CO$

$Si + C \longrightarrow SiC$

(xi) $6S_2C1_2 + I6NH_3 \xrightarrow{CCl_4} \underset{\text{Tetrasulphur-tetranitride}}{S_4N_4} + 8S + 12NH_4C1$

(xii) $2BrO_3^- + 10Cl^- + 12H^+ \longrightarrow Br_2 + 5Cl_2 + 6H_2O$

(xiii) $NCl_3 + 4H_2O \longrightarrow NH_4OH + 3HOCl$

(xiv) $BrO_3^- + 5Br^- + 6H^+ \longrightarrow 3Br_2 + 3H_2O$

(xv) $B_3N_3H_6 + 9H_2O \longrightarrow 3NH_3 + 3H_3BO_3 + 3H_2$

(xvi) $3SiF_4 + 4H_2O \longrightarrow H_4SiO_4 + 2H_2SiF_6$

(xvii) $S_4N_4 + 6OH^- + 3H_2O \longrightarrow S_2O_3^{2-} + 2SO_3^{2-} + 4NH_3$

The reaction steps and condition for the following conversions:

(i) PCl_5 to poly-dichlorophosphazene

(ii) CO to Urea

(i) ***Conversion of PCl_5 to Poly-dichlorophosphazene:*** It takes place at 120° - 150°C in an inert solvent like tetrachloroethane.

$$xPCl_5 + xNH_4Cl \longrightarrow (NPCl_2)_x + 4x\ HCl$$

The reaction steps may be

(1) $NH_4Cl + PCl_5 \longrightarrow \underset{\text{unstable}}{NH_4PCl_6}$

(2) $NH_4PCl_6 \longrightarrow HNPCl_3 + 3HCl$

(3) $xHNPCl_3 \xrightarrow{\text{condensation}} (NPCl_2)_x + xHCl$

(ii) ***Conversion of CO to Urea:*** It takes place in following steps:

$$CO + Cl_2 \longrightarrow \underset{\text{Carbonylchloride (phosgene)}}{COCl_2}$$

$$Cl_2C{=}O + 2NH_3 \xrightarrow{\text{gas phase}} (NH_2)_2C{=}O + 2HCl$$

or $$CO + \frac{1}{2}O_2 \longrightarrow CO_2$$

$$CO_2 + 2NH_3 \xrightarrow[200\,atm]{185°C} \underset{\text{Ammonium carbamate}}{H_2N-\overset{\overset{\displaystyle O}{\|}}{C}-ONH_4}$$

$$H_2N-\overset{\overset{\displaystyle O}{\|}}{C}-ONH_4 \xrightarrow{-H_2O} \underset{\text{Urea}}{NH_2CONH_2}$$

The reactions supplying the missing reactions and products:

(i) $3BC1_3 + 3NH_4C1 \longrightarrow$

(ii) $A1_2(CH_3)_6 + 2H_2O \longrightarrow$

(in) $NO + O_3 \longrightarrow$

(iv) $n[(CH_3)_2\ SiO_4] + (CH_3)_3\ SiOSi\ (CH_3)_3 \xrightarrow{H_2SO_4} A$

(v) $[3Ca_3(PO_4)_2]\ CaF_2 + 7H_2SO_4 \longrightarrow$

(vi) $PtF_6 + O_2 \longrightarrow$

(vii) $SbF_5 + BrF_3 \longrightarrow$

(viii) $PI_3 + 3H_2O \longrightarrow$

(i) $3BC1_3 + 3NH_4C1 \longrightarrow \underset{\text{B-Trichlorobenzene}}{Cl_3B_3N_3H_3} + 9HCl$

(ii) $A1_2(CH_3)_6 + 2H_2O \longrightarrow 2Al(OH)_3 + CH_4$

Correct balanced equation is

$A1_2(CH_3)_6 + 6H_2O \longrightarrow 2Al(OH)_3 + 6CH_4$

(iii) $NO + O_3 \longrightarrow NO_2 + O_2$

(iv) $n[(CH_3)_2SiO_4] + (CH_3)_3\ SiOSi\ (CH_3)_3 \xrightarrow{H_2SO_4}$

$(CH_3)_3SiO[Si(CH_3)_2O]Si(CH_3)_3$

(v) $[3Ca_3(PO_4)_2]\ CaF_2 + 7H_2SO_4$

$\underbrace{3Ca(H_2PO_4)_2 + 7CaSO_4}_{\text{Superphosphate}} + 2HF$

(vi) $PtF_6 + O_2 \longrightarrow O_2^+\ [PtF_6]^-$

(vii) $SbF_5 + BrF_3 \longrightarrow SbF_3 + BrF_5$

(viii) $PI_3 + 3H_2O \longrightarrow H_3PO_3 + 3HI$

The balanced equation for the reactions of the following with water under ambient conditions :

(i) $PC1_3$ (ii) $NC1_3$ (iii) BrF_3 (iv) Al_4C_3 (v) SF_6

(i) $PCI_3 + 3HOH \longrightarrow \underset{\text{Phosphorus acid}}{H_3PO_3} + 3HC1$

(ii) $NCl_3 + 4H_2O \longrightarrow NH_4OH + \underset{\text{Hypochlorous acid}}{3HOC1}$

(iii) $2BrF_3 + 3H_2O \longrightarrow \underset{\text{Bromic acid}}{HBrO_3} + BrF + 6HF$

(iv) $Al_4C_3 + 12H_2O \longrightarrow 4A1(OH)_3 + \underset{\text{Methane}}{3CH_4}$

(v) $SF_{6(g)} + H_2O_{(g)} \longrightarrow SO_{3(g)} + 6HF; \Delta G = -200$ kJ

Explained:

(i) B—F bond is larger in BF_4^- than in BF_3 molecule.

(ii) The order of Lewis acid strength of different halides of boron is

$$BF_3 < BC1_3 < BBr_3$$

or, BBr_3 is a stronger Lewi's acid than BF_3.

(iii) $N(CH_3)_3$ is pyramidal in shape while $N(SiH_3)_3$ has planar triangular arrangement.

or

$(SiH_3)_3N$ is a weaker base than $(CH_3)_3N$.

(iv) Dipole moment of NH_3 molecule is larger than that of NF_3.

(v) PF_3 can act as donor molecule while NF_3 show little tendency to act as donor.

(vi) $C1F_3$ exists whereas $FC1_3$ does not.

(vii) Ba(OH)$_2$ is fairly soluble in water but Mg (OH)$_2$ is not.

(viii) Dipole moment of CH$_3$CI is greater than that of CH$_3$F.

(ix) HClO$_4$ is an acid and an oxidising agent whereas H$_2$C$_2$O$_4$ is an acid and a reducing agent.

(i) B—F bond is larger in BF_4^- than in BF_3 molecule. This is due to the fact that in BF_4^-, boron atom is sp^3 hybridised while in BF_3, boron atom is sp^2 hybridised and bond length decreases with increase in *s*-character since *s*-orbital is samller than a *p*-orbital. In case of sp^3 hybridisation 25 per cent *s*-character is there while in sp^2 hybridisation 33-3 per cent *s*-character is there.

(ii) The order of Lewis acid strength of different halides of boron is

BBr$_3$ is a stronger'Lewis acid than BF$_3$.

This order is just reversed than that expected on the basis of electronegativity values of halogens. The electron density from the filled orbitals on halogens is transferred to the empty orbital present on the boron atom. This is known as back donation. This is explained on the basis of overlapping of the 2*p*-filled orbital of halogens sidewise with the empty 2*p*-orbital of boron atom forming *pπ-pπ* back bonding

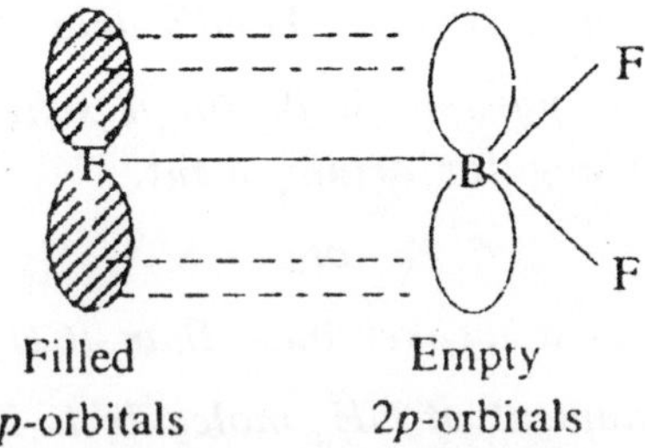

Filled 2*p*-orbitals   Empty 2*p*-orbitals

Due to back bonding, the electron deficiency in the boron atom is decreased, consequently, its Lewis character or electron accepting tendency decreases. Since back bonding is maximum in case of fluorine because of its small size, BF$_3$ is least acidic. The tendency of back bonding decreases as $BF_3 > BC1_3 > BBr_3$, the acidic character falls as $BBr_3 > BC1_3 > BF_3$.

The tendency of back bonding decreases as $BF_3 > BC1_3 > BBr_3$ because the overlapping of *3p* and *4p* filled orbitals of Cl and Br atoms with empty 2p-orbitals of B atoms does not take place effectively due to difference in the energy state of the orbitals involved.

(iii) The geometry around the nitrogen atom in trimethyl amine $N(CH_3)_3$ is pyramidal (*sp^3* hybridisation) due to lone pair-bond pair repulsion.

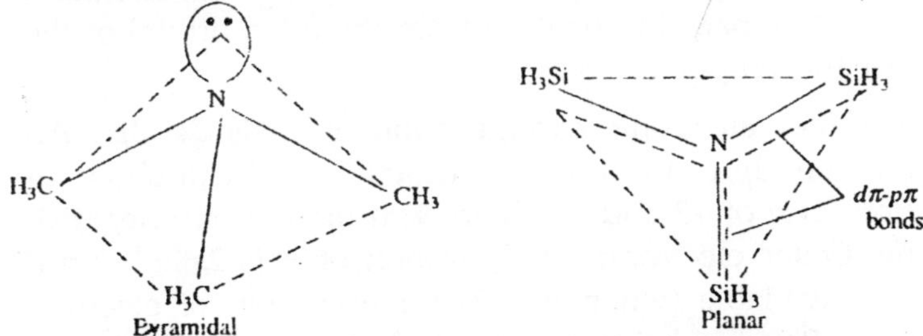

In case of similar silicon compound, $N(SiH_3)_3$ called trisilylamine, it is planar triangular arrangement of its three bonds (*sp^2* hybridisation N atom). In this case the lone pair on nitrogen present in *2p* orbital is transferred to the empty *d*-orbital of silicon forming *dπ-pπ* bond. Moreover, this makes (SiH_3) N a weaker base than $(CH_3)_3$ N.

(iv) The dipole moment of NH_3 is more than that of NF_3. This is due to different directions of the bond moments of N—H and N—F bonds. In the first case N is more electronegative but in second case F is more electronegative as shown below

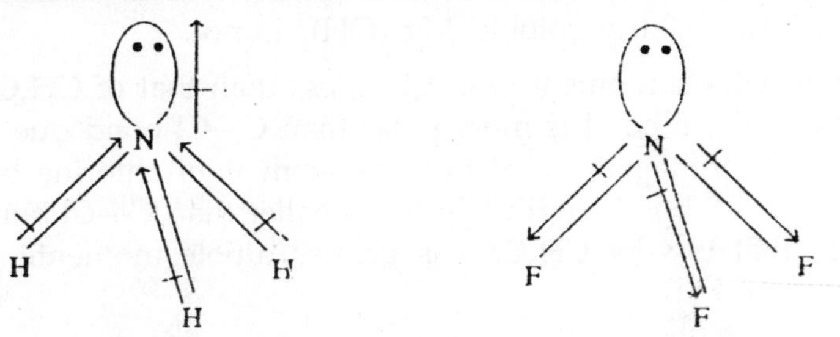

Thus in NH_3, the dipole moments of N—H bonds are in the same direction as that of the lone pair but in NF_3, the dipole moments at N—F oppose that of the lone pair.

(v) Both NF_3 and PF_3 are weak Lewis bases or donors due to the presence of a lone pair of electron on the central atom. PF_3 acts as a donor while NH_3 shows little tendency to act as donor is because phosphorus is less electronegative than nitrogen and phosphorus has a larger size than nitrogen. Due to greater electronegativity N cannot easily part with the lone pair of electrons.

In metal complexes, the *a* donation by PF_3 to metal atom is stabilised by back 7t donation from the filled orbital of the metal to the empty orbital of F.

(vi) Outer electronic configuration of Cl is $3s^2$ $3p^5$. An electron from $3p$ can jump to $3d$ orbitals, so it can show an oxidation state of +3 and combine with more electronegative fluorine. Outer electronic configuration of F is $2s^22p^5$. No *d* orbital is available (when $n = 2$) for excitation of electron. Moreover, fluorine being the most electronegative element, it shows .an oxidation state of – 1 only.

(vii) $Ba(OH)_2$ is fairly soluble in water but $Mg(OH)_2$ is not. Both Ba and Mg are the members of IIA or 2 group (alkaline earth metals). Thus in general the solubility of the alkaline earth metal hydroxides in water increases with increase in atomic number down the group. This is due to the fact tot lattice energy decreases down the group due to increase in size of alkaline earth metal cation whereas the hydration energy of the cations remains almost unchanged. Thus AH solution becomes more negative as we move from $Be(OH)_2$ to $Ba(OH)_2$ which accounts for increase in solubility or $Ba(OH)_2$ is fairly soluble Mg $(OH)_2$ is not.

(viii) Dipole moment of CH_3F is less than that of CH_3C1: No doubt C—F bond is more polar than C—Cl bond due to greater electronegativity of fluorine atom than chlorine but actually C—F bond length is much smaller than C—Cl bond length, that is why CH_3C1 has greater dipole moment.

(ix) $HClO_4$ is a strong acid and an oxidising agent. Whereas $H_2C_2O_4$ is also an (weak) acid but a reducing agent. Chlorine is more electronegative than carbon and hence can accept electron and get converted to Cl^-. In $HClO_4$, the chlorine is in its maximum oxidation state so cannot act as a reducing agent.

Recovery of elemental silver from silver resides from photographic processing (AgCl) is achieved by converting it into A, using common ionic compound B. The compound A upon heating decomposes to give an intermediate compound C before giving metallic silver as the end product. A, B and C by giving equations for the reaction involved, identified:

(a) When AgCl is fused with Na_2CO_3, Ag_2CO_3 is formed.

$$2AgCl + \underset{(B)}{Na_2CO_3} \longrightarrow \underset{(A)}{Ag_2CO_3} + 2NaCl$$

$$Ag_2CO_3 \xrightarrow{\text{Heat}} \underset{(C)}{Ag_2O} + CO_2$$

$$2Ag_2O_3 \xrightarrow{\text{Heat}} 4Ag + O_2$$

$AlBr_3$ dimerises to Al_2Br_6, while BCI3 is monomeric:

$AlBr_3$ in the vapour state as well as in inertorganic solvents exists as dimer. In the dimeric structure, each aluminium atom forms one coordinate bond by accepting a lone pair of electrons from bromine atom of another $AlBr_3$ molecule and thus completes an octet of electrons.

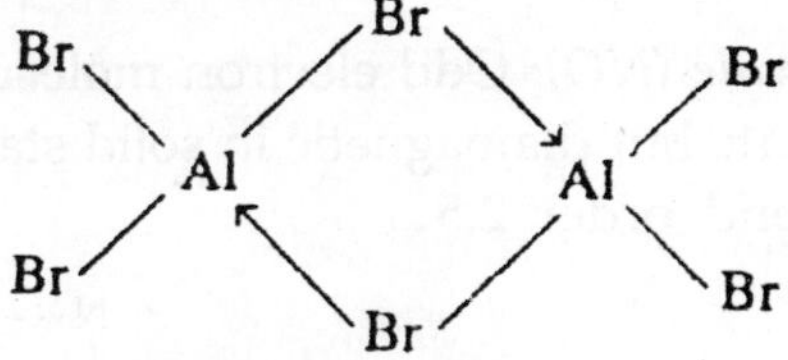

Both boron and aluminium halides are Lewis acids but only aluminium halides exist as dimer whereas boron halides exists as monomers. This is due to the reasons that boron atom is so small that it cannot accommodate four large sized halides ions around it. Moreover, in case of boron halides there is some back donation (π-back bonding) from halide to boron.

Experimentally found value of Si—F distance in SiF_4 is less than that of the theoretical value, reason:

Experimental Values of Si: F bond distance is less than theoretical values because of a dative π-back bonding arising from the presence of filled p-orbital in fluorine and empty *'d'* orbital in silicon.

HC1 is not a suitable acid medium permanganimetric titrations:

In permanganimetric titration $KMnO_4$ is always acidified with H_2SO_4 because H_2SO_4 is not self oxidising. HCl and HNO_3 are never used as $KMnO_4$ reacts with HC1 liberating chlorine ($KMnO_4$ is oxidised) liberating chlorine

$$2KMnO_4 + 16HC1 \longrightarrow 2KC1 + 2MnCl_2 + 5C1_2 + 8H_2O$$

HNO_3 is not used because it is itself an oxidising agent.

The structure of oxides of nitrogen and phosphorus.

Structure of Oxides of Nitrogen

Nitrous Oxide (N_2O): A symmetrical linear molecule, polar structure

$$N \xrightarrow{1.126Å} N \xrightarrow{1.186Å} O$$

$$N \equiv \overset{+}{N} — \overset{-}{N} \longleftrightarrow \overset{-}{N} = \overset{+}{N} = O$$

Nitric Oxide (NO): Odd electron molecule, paramagnetic in gaseous state but diamagnetic in solid state due to dimeric structure, Bond order 2.5

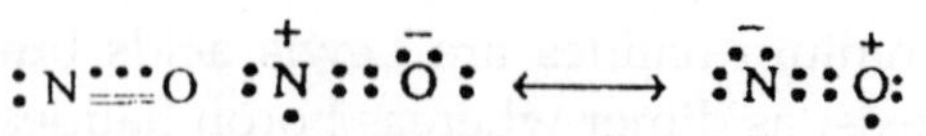

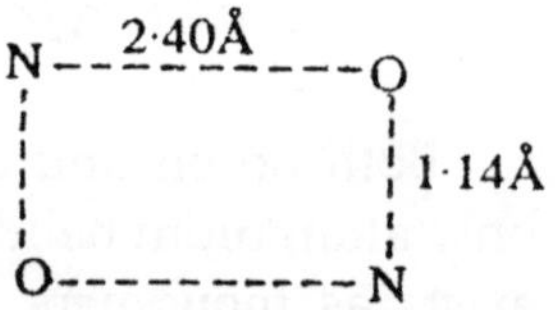

$$:N\overset{\cdot\cdot\cdot}{=\!=}O \quad :\overset{+}{\dot{N}}::\overset{-}{\ddot{O}}: \longleftrightarrow :\overset{-}{\ddot{N}}::\overset{+}{\dot{O}}:$$

N --- 2·40Å --- O
1·14Å
O ------------ N

NO: KK σ $(2s)^2$ $\overset{*}{\sigma}$ $(2s)^2$ σ $(2p_z)^2$ $\pi(2p_x)^2$ $\pi(2p_y)^2$ $\overset{*}{\pi}$ $(2p_x)^1$

Nitrogen Sesquioxide (N_2O_3)

$$NO_{(g)} + NO_{2(g)} \underset{\text{Room Temp.}}{\overset{-20°C}{\rightleftharpoons}} N_2O_{3(l)}$$

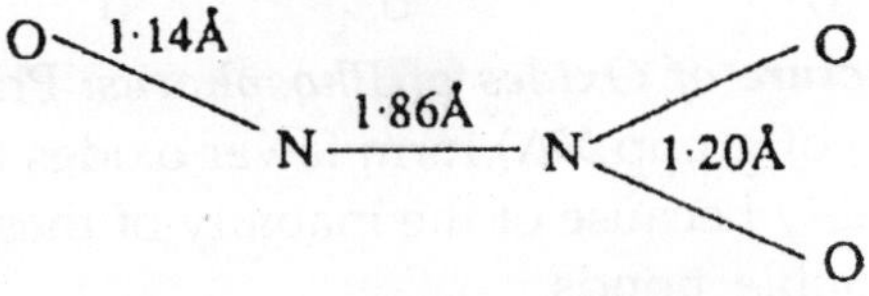

O—N 1·14Å; N—N 1·86Å; N—O 1·20Å

The oxide exists in two different forms. These may be intercon verted by irradiation with light of appropriate wavelength.

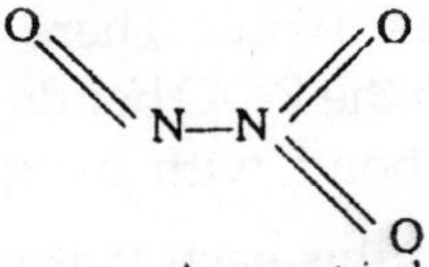

Asymmetrical 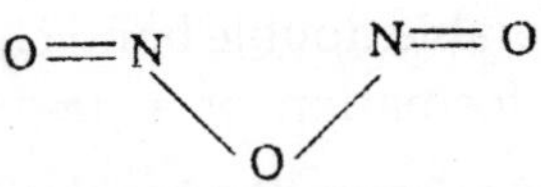 Symmetrical

Nitrogen Dioxide (NO_2) and Dinitrogen Tetraoxide (N_2O_4)

The NO_2 is obtained as a brown liquid and on cooling becomes colourless solid due to dimerisation of NO_2 into N_2O_4

$$\underset{\text{Brown paramagnetic liquid}}{NO_2} \rightleftharpoons \underset{\text{Colourless Diamagnetic solid}}{N_2O_4}$$

NO_2 is an odd electron angular molecule $O = \dot{N} \rightarrow O$

N_2O_4 has planar structure

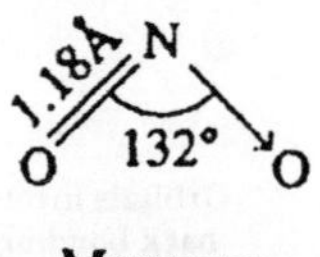

1.18Å 132°

Monomer

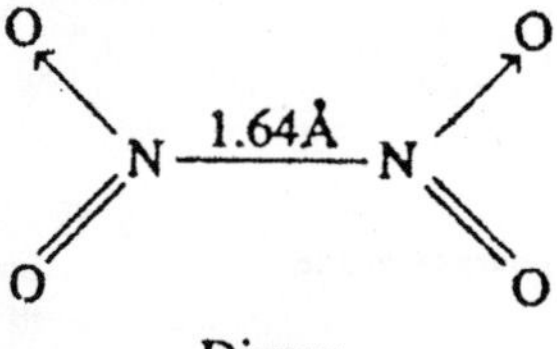

1.64Å

Dimer

Nitrogen Pentoxide (N_2O_5)

Solid N_2O_5 is ionic NO_2^+ NO_3^-. It is covalent in solution and in gas phase and has the structure.

```
O              O        O              O
 \            /          \\           ↗
  N — O — N               N — O — N
 /            \          ↙           \\
O              O        O              O
```

Structure of Oxides of Phosphorus: Phosphorus (or other elements of group VA) form fewer oxides than does nitrogen, presumably because of the inability of these elements to form *pp-pp* double bonds.

Phosphorus Trioxide (P_4O_6): It has four P atoms at the corners of a tetrahedron with six O atoms along the edges. Each P being covalently bonded to three O atoms, each of which is bonded to two phosphorus atoms. There exists a considerable double bond character in the P—O bonds because of the formation of a $p\pi$-$d\pi$ dative bond with oxygen.

Phosphorus Pentoxide (P_4O_{10}): In this each P atom forms three bonds to the oxygen atoms and also an additional coordinate bond with an oxygen atom. Terminal P—O bond is much shorter than P—O bond. It shows that there is a considerable $p\pi$-$d\pi$ back bonding because of the lateral overlap of full p-orbitals on oxygen with empty $d\pi$-orbitals on phosphorus.

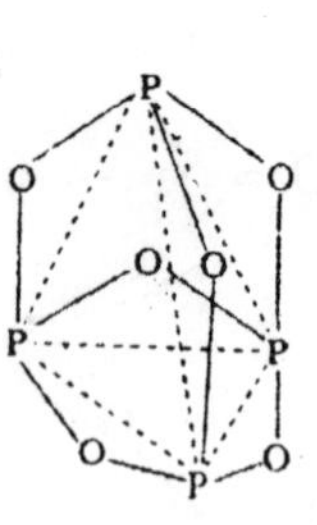

Phosphorus trioxide

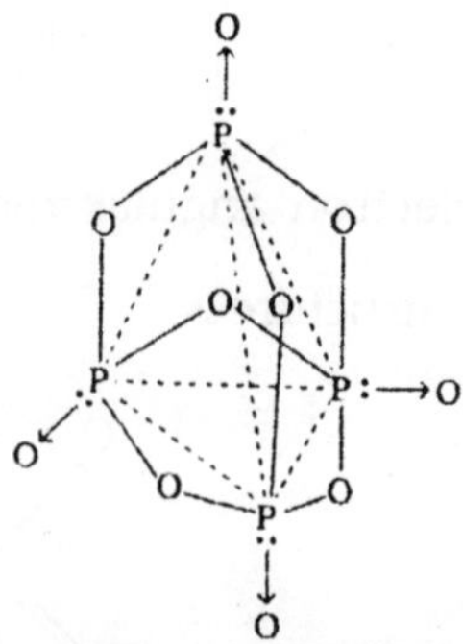

Structure of phosphorus pentoxide molecule.

Orbitals involved in back bonding

Structure of Phosphorus Halides: The halogens react .with white phosphorus to form two types of halides, PX_3 and PX_5. Iodine is an exception, since its compounds with phosphorus are PI_3 and P_2I_4. $PC1_3$ is rapidly hydrolysed to H_3PO_3.

$$PCl_3 + 3H_2O \longrightarrow H_3PO_3 + 3H^+ + 3C1^-$$

The trihalides all have pyramidal geometry similar to NH_3 and PH_3, the X—P—X bond angle is near 102°.

Pyramidal structure of PCl_3 (*sp^3* hybridisation)

Trigonal bipyramidal structure of $PC1_5$ (*sp^3d* hybridisation)

In the vapour phase the pentahalides PF_5 $PC1_5$ and PBr_5 are discrete molecules that have trigonal bipyramidal structure. Each P—Cl bond is a sigma bond formed by *sp^3d-p* overlap. The *d*-orbital used in *sp^3d* hybridisation is *$3d_z{}^2$*. In this three bond angles are of 120°, the two bond angles are 90° each. In the solid phase, however PCl_5 exists as an ionic solid consisting of $PC1_4^+$ and PCl_6^- which have the following structures,

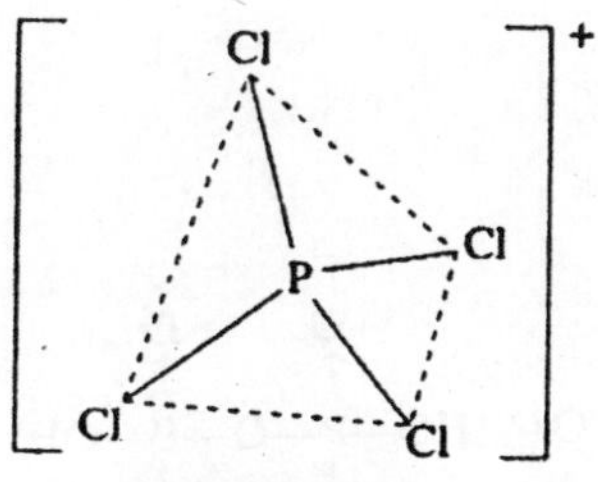

PCl_4^+ Cation
Tetrahedral structure

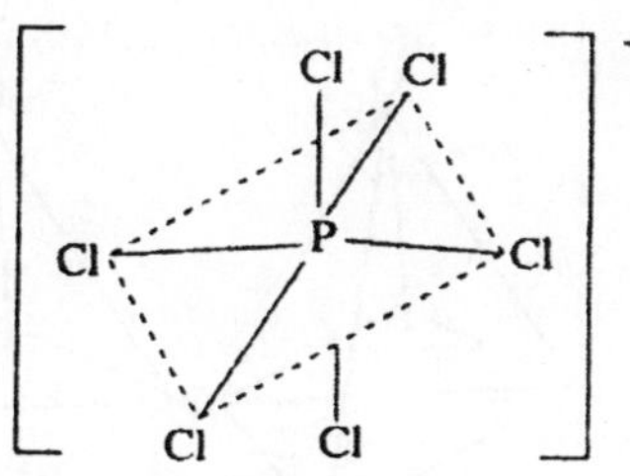

PCl_6^- anion
Octahedral structure

The phosphoryl halides (POX_3) have a slightly distorted tetrahedral structure, with X—P—X bond angles of approximately 103°.

The Structure of Compounds:

N_3H, $TeCl_4$, $H_4P_2O_7$, SO_3^{2-}, Borax

N_3H: (Hydrogenazide)—It has a bent structure.

$$\begin{array}{l} H \\ \quad \diagdown \\ \quad\; N \text{——} N \text{——} N \\ \quad\quad 1.24 \text{Å} \;\; 1.13 \text{Å} \end{array}$$

The bond angle H—N—N is 112° and two N—N bonds are of different lengths.

$TeCl_4$ (Tellurium Tetrachloride): The structure of $TeCl_4$ is established as distorted trigonal bipyramidal as shown below with one equatorial position occupied by a lone pair.

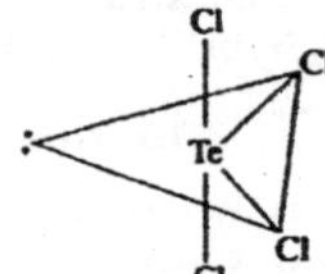

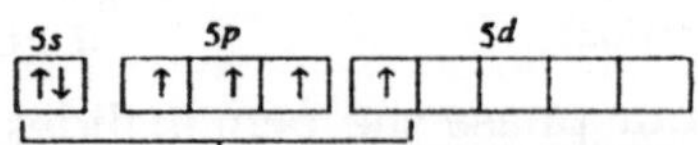

Electronic structure of tellurium atom (*Excited state*)
Four unpaired electrons form bonds with four chlorine atoms—sp^3d hybridization giving trigonal bipyramid with one position occupied by a lone pair.

$H_4P_2O_7$ (Pyro or Dipolyphosphoric Acid): It is a tetrabasic acid but gives rise to only two series of salts *e.g.*, $Na_2H_2P_2O_7$ and $Na_4P_2O_7$. It also responds to ammonium molybdate test which is characteristic for ortho-phosphates.

The acid has been assigned the following structure

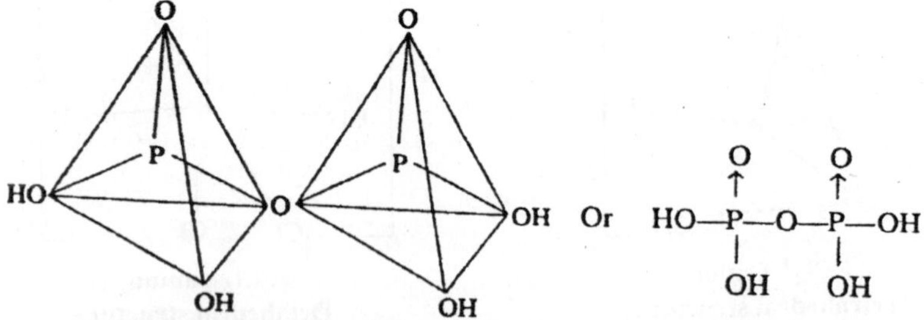

Structure of Pyrophosphoric acid

SO_3^{2-} (Sulphite Ion): The sulphite ion exists in crystals and has a pyramidal structure, that is tetrahedral with one position occupied by a lone pair. The bond angles O—S—O are slightly distorted (106°) due to the lone pair, and the bond lengths are 1-51 Å. The *n* bond is delocalised, and hence the S—O bonds have a bond order of 1.33.

Electronic structure of sulphur atom—excited state

3s	3p			3d				
↑↓	↑	↑	↑	↑				

Three unpaired electron form σ bonds forms π bond with three oxygen atoms.
Four electron pairs, hence tetrahedral with one position occupied by a lone pair.

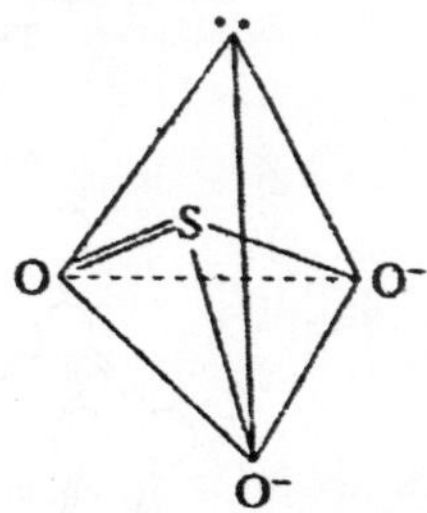

Structure of sulphite ion SO_3^{2-}

Borax ($Na_2B_4O_7$.10H2O): It is made up of two triangular and two tetrahedral units. The ion is $[B_4O_5(OH)_4]^{2-}$, remaining eight water molecules are associated with two sodium ions. Thus borax should be better formulated as $Na_2[B_4O_5(OH)_4].8x_2O$.

OH
B
O O
HO—B O B—OH
O O
B
OH

$[B_4O_5(OH)_4]^{2-}$ ion in borax

The Structures of Sulphides of Phosphorus:

Sulphides of Phosphorus: Sulphides of the formula P_4S_n (where n = 3, 5, 7, 8, 9, or 10) have been described. These may be obtained by heating red phosphorus with sulphur in appropriate amount.

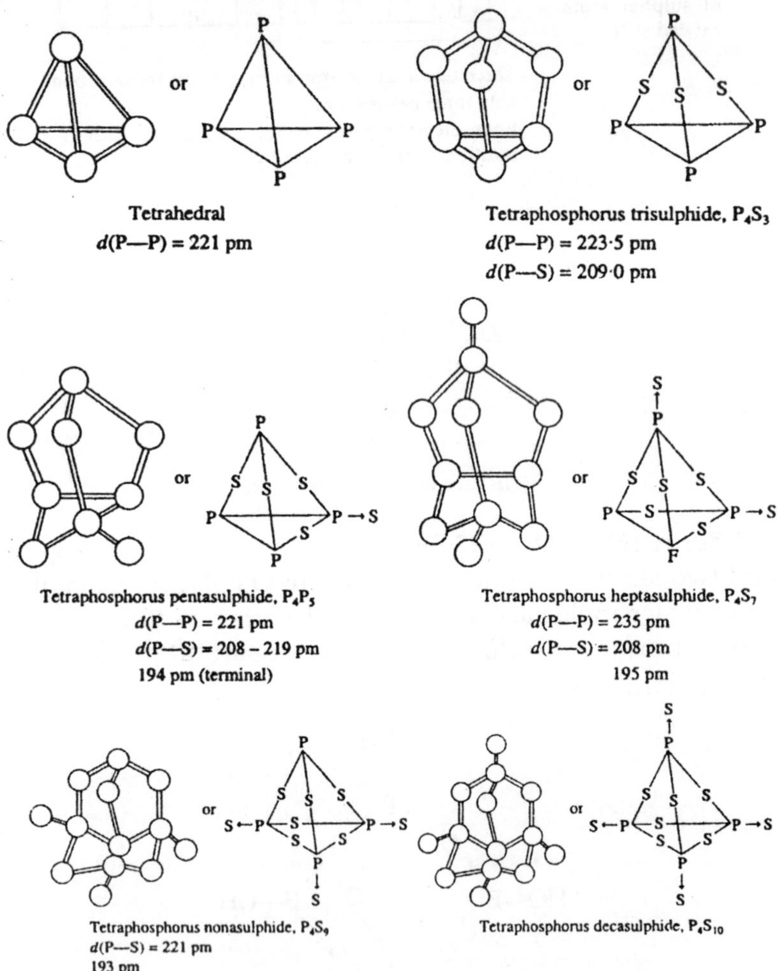

Structure of P_4 and its sulphides.

Similar to phosphorus oxides, the sulphides may be regarded as derivatives of the P_4 tetrahedral structure. There are three sulphur bridges in P_4S_3, four in P_4S_5, five in P_4S_7 and 6 in P_4S_9. Non-bridged sulphur atoms occupy terminal positions. The structure of P_4S_{10} is exactly analogous to that of P_4O_{10}.

The Structures of SO_2, SeO_2 and TeO_2, Compared:

SO_2 gas forms discrete V-shaped molecules and this structure is retained in the solid state. The bond angle is 119-30°, corresponding to a *planar trigonal structure* involving sp^2 hybridisation. Out of the three sp^2 hybrid orbitals of sulphur, one is occupied by a lone pair of electron. The remaining two half filled orbitals overlap with one of the p orbitals of each of the two oxygen atoms forming *sigma bonds*. This gives rise to planar trigonal structure, the bond angle OSO is slightly reduced from 120° to 119.5° due to repulsion between lone pair and a bond pair.

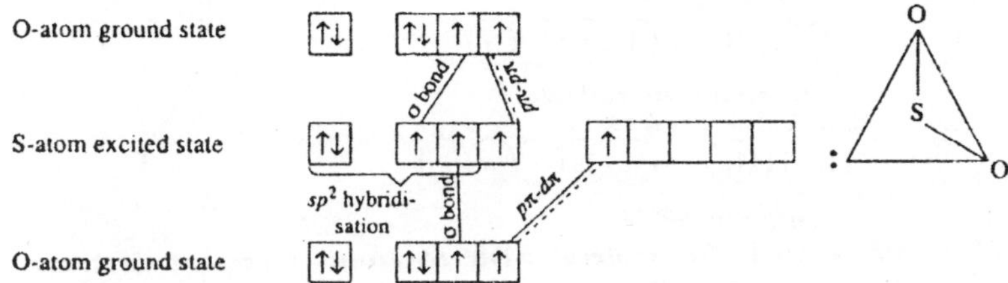

Sulphur atom has still two half-filled orbitals, *i.e.*, one p and one d-orbital. Each oxygen atom has one half-filled p-orbital. So p-orbital of one of the O-atom overlaps with p-orbital of S-atom forming $p\pi$—$p\pi$ bond. Similarly, p-orbital of the second O atom forming $p\pi$–$d\pi(3d_{xz})$ bond. Both the S—O bonds have exactly the same length because of resonance.

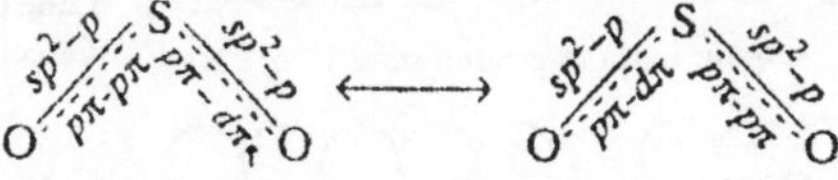

Multiple bonding explains the shorter length of S—O bond (SO_3 has planar trigonal structure involving sp^2 hybridisation of S-atom, three σ bond, one $p\pi$—$p\pi$ and $2p\pi$—$d\pi$ bonds are formed.

SeO_2 is a white volatile solid. In the gaseous state, it exists in the form of discrete molecules, just like SO_2 but in the solid state it has a polymeric structure comprising of infinite chains which are not planar.

TeO_2 is non-volatile crystalline ionic solid and occurs in two crystalline forms.

The Structure of Halides of Sulphur:

***SF_6*:** It has an octahedral structure, as shown in fig.

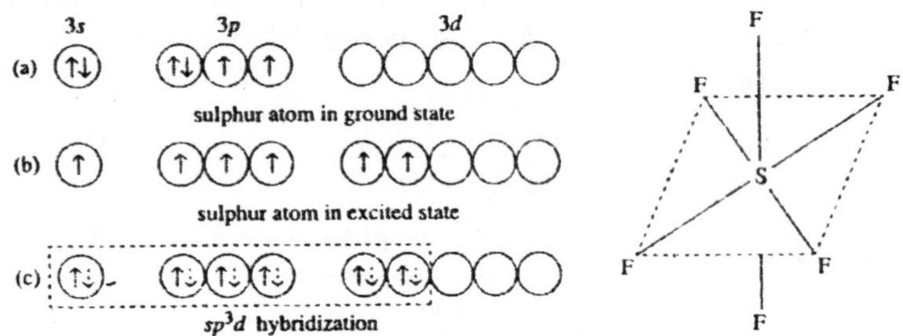

sp^3d^2 hybridisation in SF_6 molecule. Dotted arrows represent electrons supplied by fluorine atoms.

***SF_4*:** It has *trigonal bipyramidal* in structure. However, due to the presence of a lone pair of electrons, there is distortion in the molecule and the bond angles are 89° and 177° instead of normal bond angles of 90° and 180°, respectively.

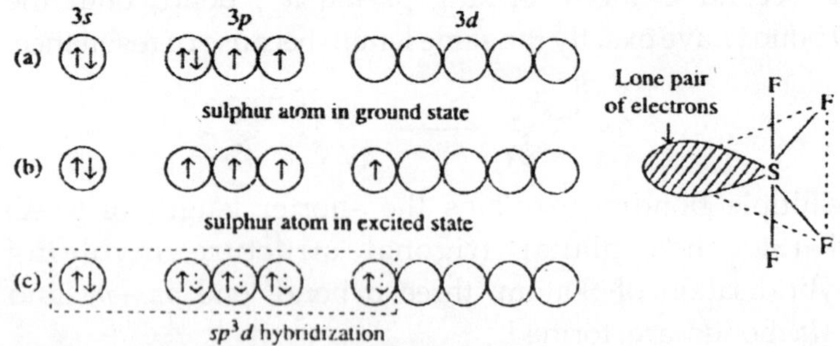

Geometry of SF_4 molecule involving sp^3d hybridization.

$SC1_2$: The formation of $SC1_2$ molecule is illustrated in figure.

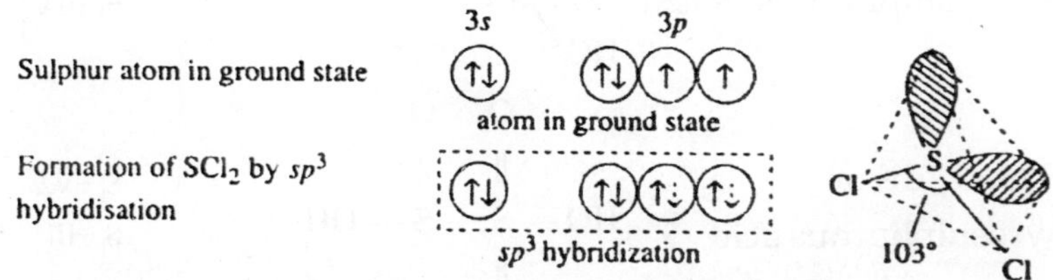

S_2F_2: Monohalides of sulphur are dimeric like S_2Cl_2, S_2F_2. The electron diffraction studies have shown that the structure of S_2F_2 is non-planar and is similar to that of H_2O_2, as shown in figure:

S_2Cl_2 has got similar structure having bond angle of 104°.

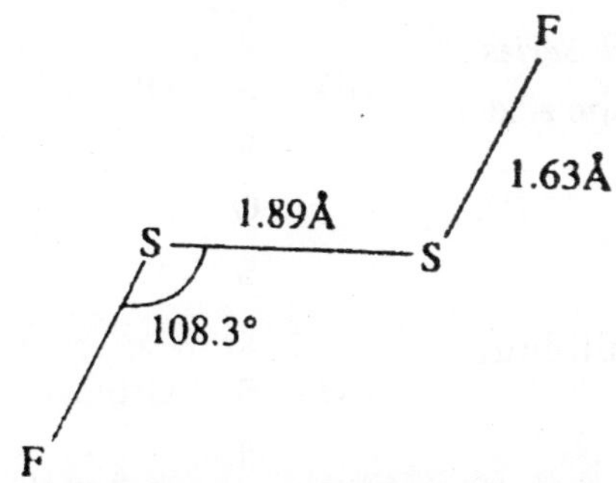

Various Types of oxoacids of Sulphur, Formed:

Oxoacids of Sulphur: Sulphur forms a number of oxoacids, many of the oxoacids of sulphur do not exist free acids, but are known as anions.

Oxoacids of sulphur may be classified into following four series :

1. Sulphurous acid series
2. Sulphuric acid series
3. Thionic acid series
4. Peroxoacid series

Sulphurous Acid Series

H_2SO_3 sulphurous acid

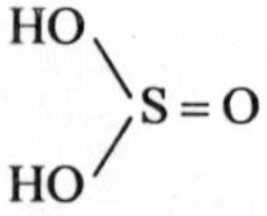

```
HO\
    S=O
HO/
```

S(+IV)

$H_2S_2O_5$ di- or pyrosulphurous acid

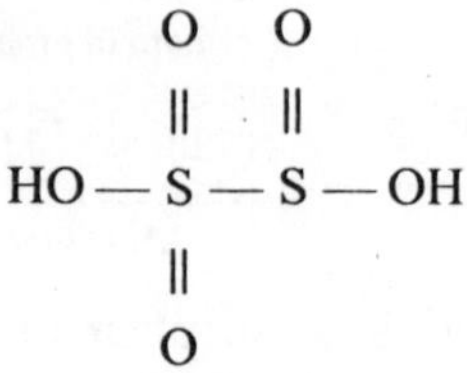

```
       O   O
       ||  ||
HO — S — S — OH
       ||
       O
```

S(+V), S(+III)

$H_2S_2O_4$ dithionous acid

```
       O   O
       ||  ||
HO — S — S — OH
```

S(+III)

Sulphuric Acid Series

H_2SO_4 sulphuric acid

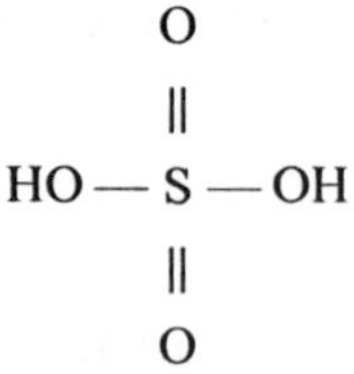

```
       O
       ||
HO — S — OH
       ||
       O
```

S(+IV)

$H_2S_2O_3$ thiosulphuric acid

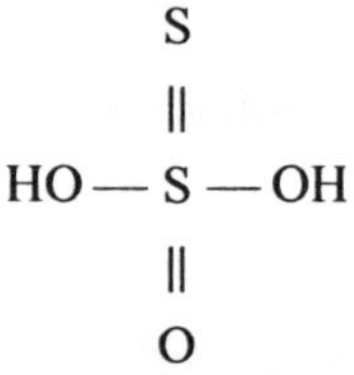

```
       S
       ||
HO — S — OH
       ||
       O
```

S(+IV), S(−II)

$H_2S_2O_7$ di- or pyrosulphuric acid

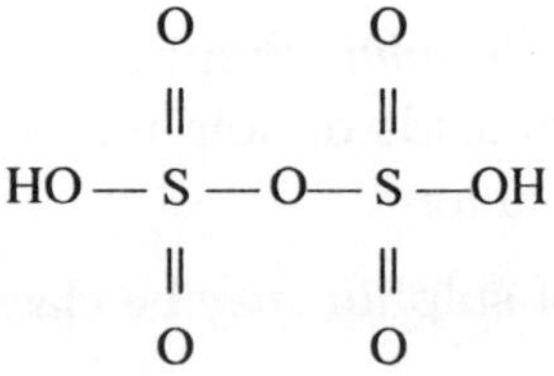

```
       O        O
       ||       ||
HO — S — O — S —OH
       ||       ||
       O        O
```

S(+VI)

Thionic Acid Series

$H_2S_2O_6$ dithionic acid

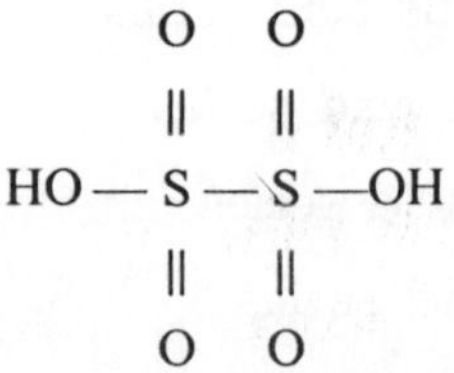

```
       O   O
       ||  ||
HO — S — S —OH
       ||  ||
       O   O
```

S(+V)

$H_2S_nO_6$ polythionic acid (n=1-12)

$$HO-\overset{\overset{O}{\|}}{\underset{\underset{O}{\|}}{S}}-(S)_n-\overset{\overset{O}{\|}}{\underset{\underset{O}{\|}}{S}}-OH$$

S(+V), S(0)

Peroxoacid Series

H_2SO_5 peroxomono-sulphuric acid

$$HO-\overset{\overset{O}{\|}}{\underset{\underset{O}{\|}}{S}}-O-OH$$

S(+VI)

$H_2S_2O_8$ peroxodi-sulphuric acid

$$HO-\overset{\overset{O}{\|}}{\underset{\underset{O}{\|}}{S}}-O-O-\overset{\overset{O}{\|}}{\underset{\underset{O}{\|}}{S}}-OH$$

S(+VI)

Common Qualities

Reactivity: Interhalogen compounds are more reactive than the halogen (except F_2). This is because the A—X bond in interhalogen is weaker than the X—X bond in the halogens. The order of reactivity of some interhalogen compounds has been found to be

$$ClF_3 > BrF_5 > IF_7 > ClF > BrF_3 > IF_5 > BrF > IF_3 > IF$$

Hydrolysis: Hydrolysis gives halide and oxohalide ions. Oxohalide ion is always formed from the larger halogen present. Oxidation state of the larger halogen does not change during hydrolysis.

$$ICl \xrightarrow{H_2O} Cl^- + OI^-$$

$$2ICl_3 + 3H_2O \longrightarrow 5HCl + HIO_3 + ICl$$

$$BrF_5 + 3H_2O \longrightarrow \underset{(F^-)}{5HF} + \underset{(BrO_3^-)}{HBrO_3}$$

$$IF_7 \xrightarrow{H_2O} IO_4^- + 7F^-$$

Halogenation: Interhalogen compounds are good halogenatig agent.

$$ClF + CsF \longrightarrow Cs^+ + ClF_2^-$$
$$2ClF + AsF_5 \longrightarrow FCl_2 + AsF_6^-$$
$$6ClF + 2Al \longrightarrow 2AlF_3 + 3Cl_2$$
$$6ClF + U \longrightarrow UF_6 + 3Cl_2$$
$$6ClF + S \longrightarrow SF_6 + 3Cl_2$$
$$4ClF_3 + 6MgO \longrightarrow 6MgF_2 + 2Cl_2 + 3O_2$$
$$4ClF_3 + 2Al_2O_3 \longrightarrow 2AlF_3 + 2Cl_2 + 3O_2$$
$$2ClF_3 + 2Al_2O_3 \longrightarrow 2AgF_2 + Cl_2 + 2ClF$$
$$2ClF_3 + 2NH_3 \longrightarrow 6HF + Cl_2 + 2ClF$$
$$2ClF_3 + 2NH_3 \longrightarrow 6HF + Cl_2 + N_2$$
$$ClF_3 + BF_3 \longrightarrow [ClF_2]^+ [BF_4]^-$$
$$ClF_3 + SbF_5 \longrightarrow [ClF_2] + [SbF_6]^-$$
$$ClF_3 + PtF_5 \longrightarrow [ClF_2] + [PtF_6]^-$$
$$ClF_3 + UF_4 \longrightarrow UF_6 + ClF$$
$$4ClF_3 + 3Pu \longrightarrow 3PuF_4 + 2Cl_2$$
$$4ClF_3 + 3N_2H_4 \longrightarrow 12HF + 3N_2 + 2Cl_2$$
$$4BrF_3 + SiO_2 \longrightarrow 3SiF_4 + 2Br_2 + 3O_2$$

(B_2O_3, As_2O_5, I_2O_5, CuO, TiO_2 react similarly)

$$2CiF + AsF_5 \longrightarrow FCl_2 + AsF_6^-$$
$$ClF_3 + AsF_5 \longrightarrow ClF_2 + AsF_6^-$$
$$ClF_3 + CsF \longrightarrow Cs_2 + ClF_4^-$$
$$2BrF_5 + SiO_2 \longrightarrow SiF_4 + 2BrF_3 + O_2$$
$$BrF_5 + CsF \longrightarrow Cs + [BrF_6]^-$$
$$IF_5 + KI \longrightarrow K^+ + [^-IF_6]^-$$
$$2IF_7 + SiO_2 \longrightarrow 2IOF_5 + SiF_4$$
$$IF_7 + CsF \longrightarrow Cs^+ + [IF_8]^-$$

Self Ionisation: In liquid state many interhalogen have an appreciable electrical conductivity due to self ionisation.

$$2ICl \rightleftharpoons I^{+} + [ICl_2]^{-}$$

$$2BrF_3 \rightleftharpoons [BrF_2]^{+} + [BrF_4]^{-}$$

$$2IF_5 \rightleftharpoons [IF_4]^{+} + [IF_6]^{-}$$

The Structure of Following Compounds:

ClF_3, I_2Cl_6, IF_5, IF_7.

The structure of these compounds can be represented as:

ClF_3: The molecule is T-shaped (structure 3) with bond angle of 87°40'. The distortion from 90° is because of repulsion^between the lone pairs. Equatorial bonds (those in triangle) are different from apical bonds (those pointing ups and down)

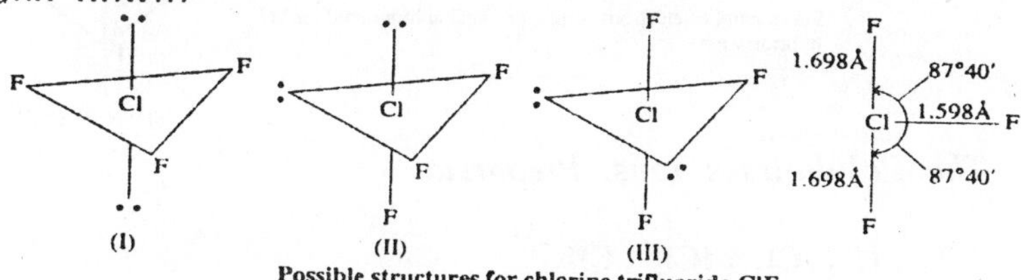

Possible structures for chlorine trifluoride ClF_3

The Structure of BrF_3 is also T-shapes:

Electronic structure of central chlorine atoms in ClF_3 (excited state)

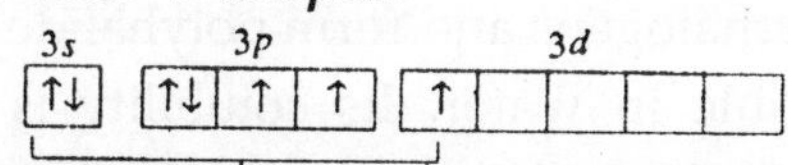

Three unpaired electrons form bonds with three fluorine atoms. sp^3d hybridisation trigonal bipyramid with two positions occupied by lone pairs.

I_2Cl_6 I_2Cl_6: In the solid state, two T-shaped ICl_3 molecules join together, forming a planar dimeric molecule $(ICl_3)_2$ or I_2Cl_6. ICl_3 gas decomposes into ICl and- Cl_2, so its structure is not known. The liquid undergoes self ionisation as

Structure of I_2Cl_6

$$2ICl_3 \rightleftharpoons ICl_2^{+} + ICl_4^{-}$$

***IF_5*:** The IF_5 (or AX_5) compound has structure based on a square pyramid, that is octahedral with one position occupied by lone pair. The central atom is displaced slightly below the plane.

Electronic structure of iodine atom— excited state

5s	5p			5d				
↑↓	↑	↑	↑	↑	↑			

Five unpaired electrons form bonds with five fluorine atoms, plus one lone pair giving a total of six electron pairs structure is octahedral (sp^3d^2 hybridisation) with one position occupied by a lone pair (alternatively described as square based pyramid)

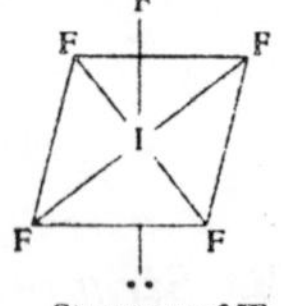

Structure of IF_5.

***IF_7*:** The structure of IF_7 is usual a pentagonal bypyramid. It is probably the only known example of a non-transition element using three *d* orbitals for bonding.

Electronic structure of iodine atom— excited state

5s	5p			5d				
↑	↑	↑	↑	↑	↑	↑		

Seven unpaired electrons form bonds with seven fluorine atoms,
Seven pairs of electrons form a pentagonal bipyramid (sp^3d^3 hybridisation)

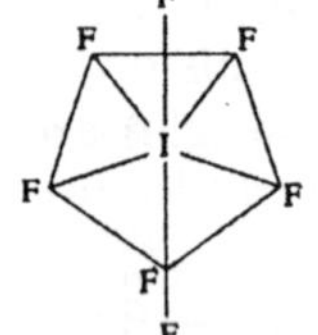

Structure of IF_7

The Polyhalides Ions, Prepared:

$$I_3^-, I_5^{3-}, ICl_2^-, ICl_4^-, ClF_2^+$$

Halide ions often react with molecules of halogens or interhalogens and form polyhalide ions. Iodine is only slightly soluble in water. Its solubility is greatly increased if same iodide ions are present in the solution. The increase in solubility is due to the forma'tion of polyhalide ion, in this case triiodide ion I_3 ~. This is stable both in aqueous solution and in ionic crystals.

$$I_2 + I^- \longrightarrow I_3^-$$

Br_3^- and Cl_3^- ions are less stable. No F_3^- compounds are known presumably because fluorine has no available *d*-orbitals and, therefore, cannot expand its octet.

$$Cl^- + Cl_2 \longrightarrow Cl_3^-$$

Many polyhalides are known which contain two or three different halogens.

$$IC1 + KC1 \longrightarrow K + [IC1_3]^-$$

$$IC1_3 + KC1 \longrightarrow K^+[IC1_4]^-$$

$$IF_5 + CsF \longrightarrow Cs + [IF_6]^-$$

$$2C1F + AsF_5 \longrightarrow [FC1_2]^- [AsF_6]^-$$

$$BrF_3 + C1F_3 \longrightarrow [ClF_2] + [BrF_4]^-$$

Structure and Shape of I_3^- Ion: I_3^- ion has linear shape as shown in Fig.

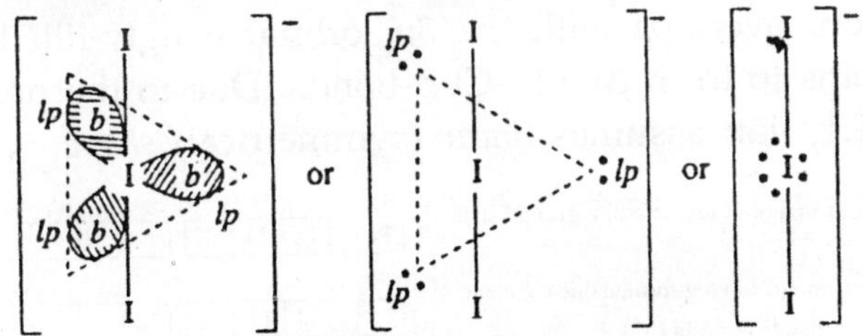

Linear shape of I_3^- ion with *bps* = 2 and *lps* = 3.

In some cases I—I—I bond angle in I_3^- ion is not equal to 180° and the two I—I bond lengths are unequal (unsymmetrical structure) as shown below in some ionic salts.

I 2·90Å I 2·93Å I, 176°
$[(C_6H_5)_4As]^+ I_3^-$

I 3·10Å I 2·82Å I, 177°
$[NH_4]^+ I_3^-$

I 2·83Å I 3·04Å I, 177°
$Cs^+ I_3^-$

Structure of I_5^- Ion: The structure of pentaiodide in $(CH_3)_4NI_5$ is quite different from those of other polyhalides containing five halogens. It is planar and V-shaped.

I 3.17Å I 2.82Å I
3.17Å, 94°
80°
2.82Å
I

Planar V-shaped geometry f I_5^- ion in $[(CH_3)_4N^+I_5^-]$

Structure and Shape of $IC1_2^-$ ***Ion:*** This ion has symmetrical linear shape which results from sp^3d hybridisation of I-atom (central atom). Due to the presence of one unit of negative charge on ICI_2^- ion, I-atom (central atom) can be regarded as having eight electrons (instead of seven) in its valence-shell. The Lewis structure of $IC1_2^-$ ion, $\left[\text{Cl—}\ddot{\underset{\cdots}{\text{I}}}\text{—Cl}\right]^-$ shows that I-atom is surrounded by two σ-*bps* and the remaining six electrons remain as three *lps* on I-atom. Thus, since σ-*bps* + *Ips* = 2 + 3 = 5, I-atom is sp^3d hybridised in $IC1_2^-$ ion.

Now the two axial hybrid orbitals, each of which has one electron, overlaps with the $3p_x$ orbital (singly-filled) of two Cl-atoms to form two I—Cl σ-bonds. Due to the presence of *Ips*, $IC1_2^-$ ion assumes *linear* (symmetrical) *shape*

	5s	5p			5d				
Electronic structure of iodine atom–ground state	↑↓	↑↓	↑↓	↑					
Structure of iodine having formed one covalent and one coordinate bond in $[IC1_2]^-$.	↑↓	↑↓	↑↓	↑↓	↑↓				

sp^3d hybridisations

Five electrons pairs—trigonal bipramid with thre positions occupied by lone pairs

The three lone pairs of electrons *(Ips)* occupy the basal positions of the trigonal bipyramid.

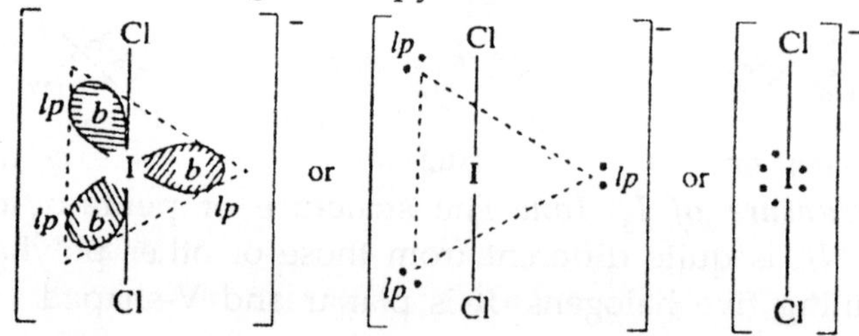

Linear (symmetrical) shape of ICl_2^- ion with *bps* = 2 and *lps* = 3.

Structure and Shape of $IC1_4^-$ ***Ion:*** In this ion I-atom is the central atom. Due to the presence of one unit of negative charge on $IC1_4^-$ ion, I-atom (central atom) can be regarded as having eight electrons (instead of seven) in its valence-shell.

The Lewis structure of this ion, $\left[\begin{matrix} Cl & & Cl \\ & \ddot{I} & \\ Cl & \cdot\cdot & Cl \end{matrix}\right]^-$ shows that I-atom is surrounded by four σ-*bps* and two *Ips*. Thus since σ-*bps* + *Ips* =4 + 2 = 6, I-atom is assumed to be sp^3d^2 hybridised. Out of the six hybrid orbitals two axial hybrid orbitals contain two *Ips*, because in order to have minimum lone pair repulsion, the lone pairs must be as far from each other as possible. Other four hybrid orbitals (equatorial hybrid orbitals) which are singly-filled overlap with the singly-filled $2p_x$ orbitals on four Cl-atoms and form four I—Cl a-bonds.

Although the spatial arrangement of six electron pairs round I-atom is octahedral, due to the presence of two lone pairs of electrons in the axial hybrid orbitals, the shape of $IC1_4^-$ ion gets distorted and becomes *square planar* as shown in Fig.

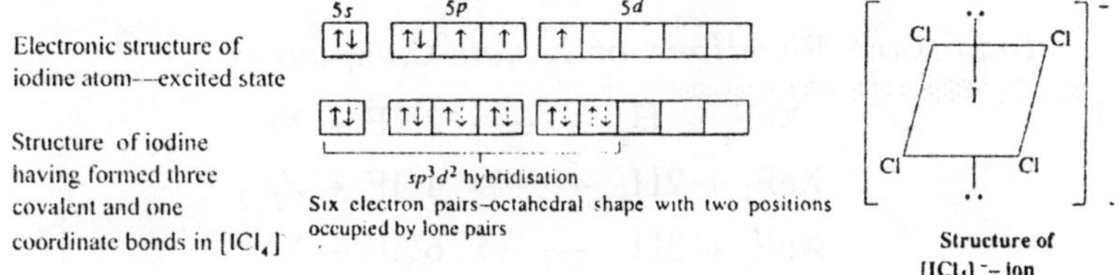

Structure of $[ICl_4]^-$ ion

(BrF_4^- also has similar structure and same explanation)

Structure and Shape of $\boldsymbol{C1F_2^+}$: In this ion Cl-atom is the central atom. Due to the presence of one unit of positive charge on C1F2⁺ ion, Cl-atom (central atom) can be regarded as having six electrons (instead of seven) in its valence-shell.

The Lewis structure of $C1F_2^+$ ion, $\left[F—\ddot{\underset{\cdot\cdot}{C}}1—F\right]^+$ shows that Cl-atom has σ-*bps* = 2 and *Ips* = 2, and, since σ-*bps* + *Ips* = 4, Cl-atom is supposed to be sp^3 hybridised.

Two sp^3 hybrid orbitals have *lps* while each of the remaining two sp^3 hybrid orbitals has one electron. These two singly-filled sp^3 hybrid orbitals make a *head-to-head* (linear) overlap with singly-filled $2p_x$ orbitals on each of the two F-atoms and form two sp^3 (Cl)—$2p_x$ (F) σ-bonds.

Although the spatial arrangement of four electron pairs round Cl-atom is tetrahedral, due to the presence of two lone pairs of electrons, the shape of ClF_2^+ ion gets distorted and becomes *angular or V-shape* as shown in Fig.

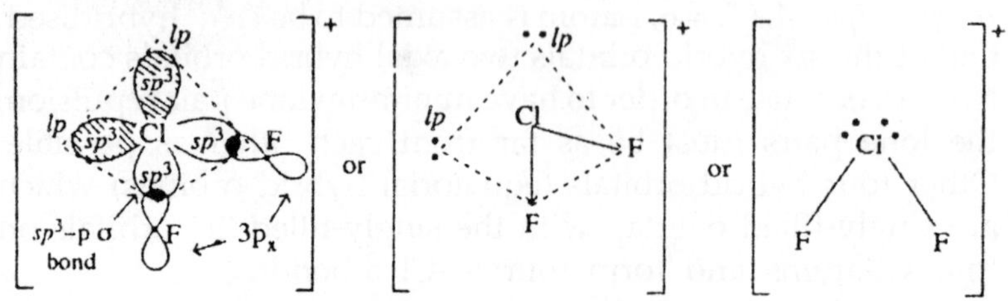

Angular or V-shape of ClF_2^+ ion (σ-*bps* = 2, *lps* = 2)

The Cl—F bond length and Cl—F—F bond angle found in ClF_2^+ ion depends on the nature of the anion attached to this cation.

(The ICl_2^+ ion also has similar structure and shape.)

Important Reactions of Xenon Compounds:

1.
$$XeF_2 + H_2 \longrightarrow 2HF + Xe$$
$$XeF_4 + 2H_2 \longrightarrow 4HF + Xe$$
$$XeF_6 + 3H_2 \longrightarrow 6HF + Xe$$

2.
$$XeF_2 + 2HCl \longrightarrow 2HF + Xe + Cl_2$$
$$XeF_4 + 4KI \longrightarrow 4KF + Xe + 2I_2$$
$$SO_4^{2-} + XeF_2 + Ce_2^{III}(SO_4)_3 \longrightarrow 2Ce^{IV}(SO_4)_2 + Xe + F_2$$

(In above reactions xenon fluoride oxidises Cl^- to Cl_2, I^- to I_2 and Ce (III) to Ce (IV).

3.
$$XeF_4 + 2SF_4 \longrightarrow Xe + 2SF_6$$
$$XeF_4 + Pt \longrightarrow Xe + PtF_6$$
$$XeF_4 + C_6H_6 \longrightarrow Xe + C_6H_5F + HF$$

(XeF_4 is used as fluorinating agent.)

4.
$$2XeF_2 + 2H_2O \longrightarrow 2Xe + 4HF + O_2$$
$$3XeF_4 + 6H_2O \longrightarrow 2Xe + XeO_3 + 1\frac{1}{2}O_2$$

$$XeF_6 + 2H_2O \longrightarrow XeOF_4 + 2HF$$
$$2XeF_6 + 6H_2O \longrightarrow XeO_3 + 12HF$$

(XeO_3 is a explosive solid.)

5. $$2XeF_6 + SiO_2 \longrightarrow XeOF_4 + SiF_4$$

6. $$XeO_3 + 2XeF_6 \longrightarrow 3XeOF_4$$
$$XeO_3 + XeOF_4 \longrightarrow 2XeO_2F_2$$

7. $$XeO_3 + NaOH \longrightarrow \underset{\text{Sodium xenate}}{Na^+ [HXeO_4]^-}$$
$$2[HXeO_4]^- + 2OH \longrightarrow [XeO_6]^{4-} + Xe + O_2 + 2H_2O$$

8. Xenon fluorides act as a fluoride donor and forms complexes with covalent pentafluorides.

   XeF_2 forms complexes with PF_5, AsF_5, SbF_6, NbF_5, TaF_5, RuF_5, OsF_5, RhF_5, IrF_5 and PtF_5. These are thought to have the structure

$$XeF_2 \cdot MF_5\ [XeF]^+\ [MF_6]^-$$
$$XeF_2.2MF_5[XeF]^+\ [M_2F_{11}]^-$$
$$2XeF_2.\ MF_5[Xe_2F_3] + [MF_6]^-$$

   XeF_4 forms only a few complexes (withPF_5, As F_5 and SbF_5 only) XeF_6 forms complexes such as XeF_6. BF_3; $XeF_6.GeF_4$; XeF_6. AsF_5; XeF_6- SbF_5.

9. XeF_6 may also act as fluoride acceptor. With RbF and CsF it reacts as

$$XeF_6 + RbF \longrightarrow Rb^+ [XeF_7]^-$$

On heating XeF_7^- ion decomposes

$$2Rb^+ [XeF_7]^- \xrightarrow{50°C} XeF_6 + Rb_2\ [XeF_8]$$

Starting from xenon and fluorine and other appropriate reagents, is prepared potassium perxenate. Using VSEPR theory, the probable structure of perxenate ion, predicted:

From xenon and fluorine, potassium perxenate can be prepared as follows:

$$\underset{1\ :\ 20}{Xe + 3F_2} \longrightarrow XeF_6$$

$$XeF_6 + 3H_2O \longrightarrow XeO_3 + 16HF$$

XeO_3 is soluble in water, but does not ionise. However, in alkaline solution above pH 10-5, it forms the xenate ion

$$XeO_3 + KOH \longrightarrow K^+ [HXeO_4]^-$$

Potassium Xenate + $2H_2O$

Xenates contain Xe (+ VI) and they slowly disproportionate in solution to perxenates (which contain Xe (+ VIII) and Xe

$$K + [HXeO_4]^- + 2OH^- \longrightarrow K_4^+ [XeO_6]^{4-} + Xe + O_2$$

Potassium perxenate + $2H_2O$

K_4XeO_6

A Brief VSEPR Explanation of Structures of Xenon Fluorine Compounds:

The structure of xenon and fluorine compounds can be summarised as follows:

Compound	Structure	No. of electron pairs	No. of lone pairs	VSEPR explanations
XeF_2	Linear	5	3	5 electron pairs forms trigonal bypyramid with three lone pairs at equitorial positions
XeF_4	Square planar	6	2	6 electron pairs form an octahed-ron with two position occupied by lone pairs
XeF_6	Distorted octahedron	7	1	Pentagonal bipyramid with one lone pair
XeO_3	Pyramid	7	1	Three *n* bonds so the remaining H electron pairs form a tetrahedron with one lone pair
XeO_2F_2	Trigonal bipyramid	7	1	Two π bonds so the remaining 5 electron pairs form trigonal bipyramid with one lone pair at equitoria) position
$XeOF_4$	Square pyramidal	7	1	One π bond so the remaining 6 electron pairs form an octahedron with one position occupied by lone pair

XeO_4	Tetrahedral	8	0	Four π bonds so remaining four electron pairs form a tetrahedron
XeO_3F_2	Trigonal bipyramid	8	0	Three π bonds so remaining five electron pairs form a trigonal bipyramid
$[XeO_6]^{4-}$	Octahedral	8	0	Two π bonds so the remaining 6 electron pairs form an

The Structures of Compounds Drawn:

(i) XeF_2 **(ii) XeF_4** **(iii) XeF_6**

(iv) XeO_3 **(v) $XeOF_4$** **(vi) XeO_2F_2**

(vii) XeO_4 **(viii) $[XeO_6]^{4-}$**

XeF_2: There are 10 valence electrons (8 from valence shell of xenon $5s^2$ $5p^6$ and two form the two bonded fluorine atoms) to be filled in orbitals. These require *five orbitals* which can be formed by the hybridisation of one *5s*, three *5p* and one *5d* orbitals (*sp^3d* hybridisation). These are directed towards the five corners of a trigonal bipyramid. Two of these contain shared electrons, the other three lone pairs. For greatest stability, the shared pairs are as far apart as possible, so Xe - F bonds are at 180° to each other and the structure is *linear.*

Xenon Atom

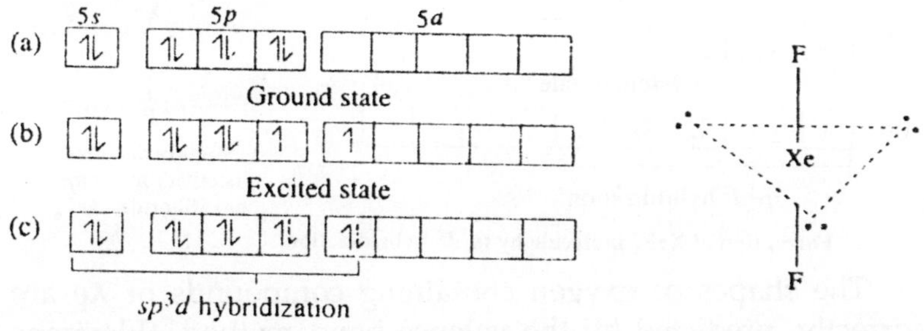

Formation of XeF_2 molecule by sp^3d hybridisation. Electrons supplied by fluorine atoms are represented by dotted arrows.

XeF_4: In a similar manner we have 8 + 4 or 12 valence electrons to accommodate (8 being electrons of the central atom and 4 being electrons from 4 fluorine atoms). These require 6 orbitals, hence *sp^d²* hybridisation. The structure is

square planar and the four fluorine atoms are in the same plane along with xenon atom. The six orbitals are directed towards the six corners of an octahedron. The two lone pairs are one above one below the plane containing Xe and 4 fluorine atoms.

Xenon Atom

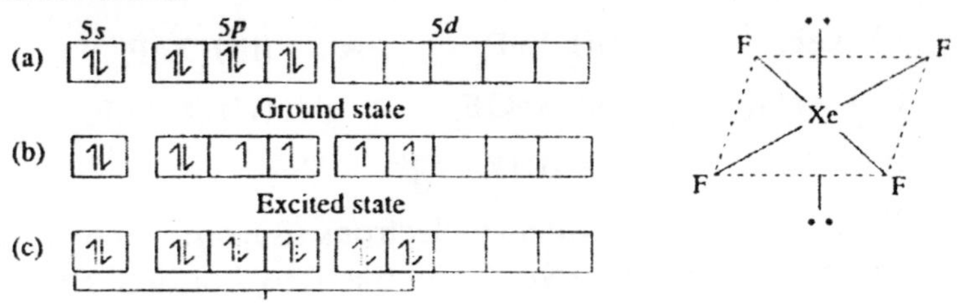

Formation of XeF_4 molecule by sp^3d^2 hybridization.

XeF_6: Here we have 8 + 6 or 14 electrons t.o accommodate. These need 7 orbitals. Hence sp^3d^3 hybridisation resulting in distorted octahedron.

Xenon Atom

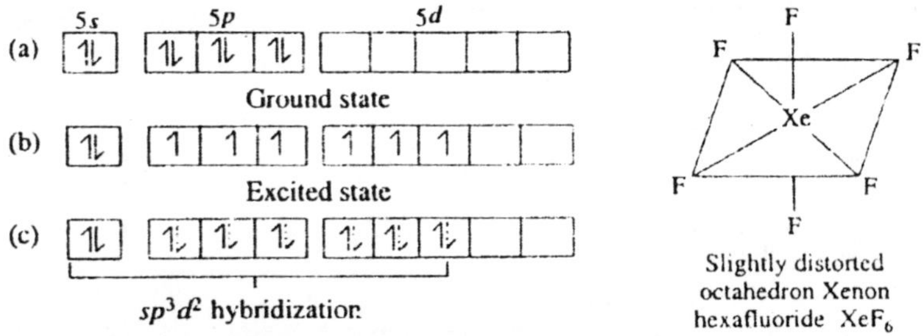

Slightly distorted octahedron Xenon hexafluoride XeF_6

Formation of XeF_6 molecule by sp^3d^2 hybridization.

The shapes of oxygen containing compounds of Xe are correctly, predicted by the valence bond method. (Electrons in π bonds (double bonds) must be subtracted before counting the number of electron pairs which determine the primary shape of the molecule.

XeC_3: Oxygen shares three of the four lone pairs of xenon. Hence only 8 valence electrons are to be accommodated and hybridisation around xenon is sp^3. As three orbitals contain bond pairs and one lone pair, so the geometry is described as trigonal pyramid. The O—Xe—O bond angle is found to be 103° and is reasonably close to the expected ≈ 109°, taking into account the greater repulsion between lone pair and bond pairs as compared to bond pair-bond pair repulsion.

Xenon Atom

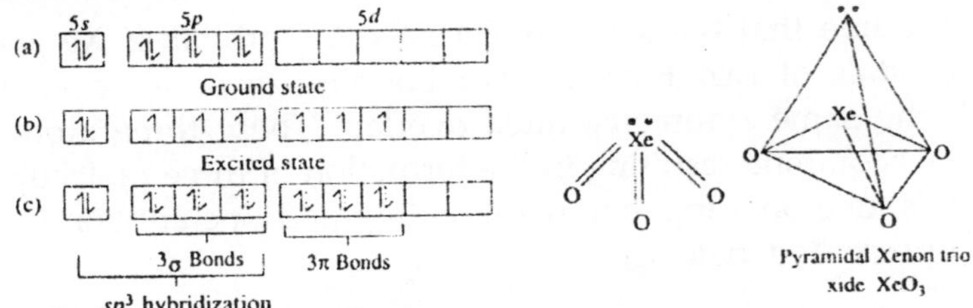

***$XeOF_4$*:** Here oxygen atom does not contribute any electrons but shares one of the lone pairs of xenon. Hence the structure is like that of XeF_4 in which one of the orbitals containing lone pairs around xenon is replaced by Xe—O bond. The structure is described as square pyramid.

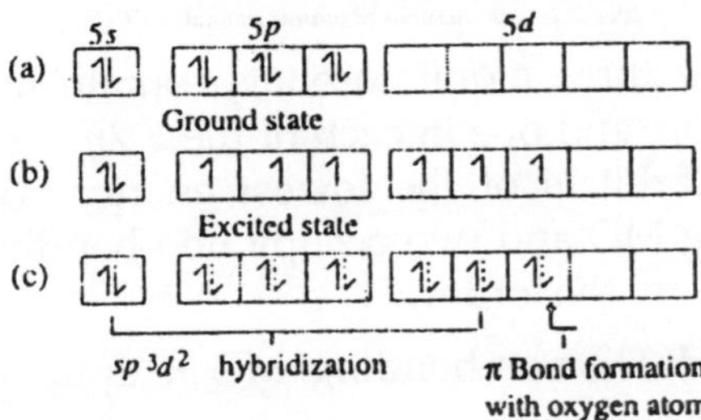

The six orbitals, as before, are arranged towards the six corners of a regular octahedron. The structure of XeO_2F_2, $[XeO_6]^{4-}$ and XeO_4 can be described on similar lines. The structures are shown below:

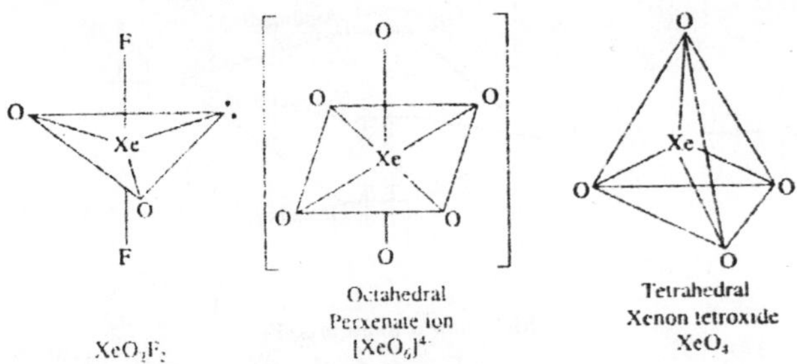

XeO_2F_2 Octahedral Perxenate ion $[XeO_6]^{4-}$ Tetrahedral Xenon tetroxide XeO_4

XeF_2: The outer electronic configuration of xenon and fluorine atoms are

Xe				F			
5s	5p			2s	$2p_x$	$2p_y$	$2p_z$
↑↓	↑↓	↑↓	↑↓	↑↓	↑↓	↑↓	↑

Assume that bonding involves the $5p_z$ orbital of Xe and $2p_z$ orbitals of two F atoms. For bonding to occur, orbitals with the same symmetry must overlap. These three atomic orbitals combine resulting in the formation of three molecular orbitals, one bonding, one non-bonding and one antibonding as represented below

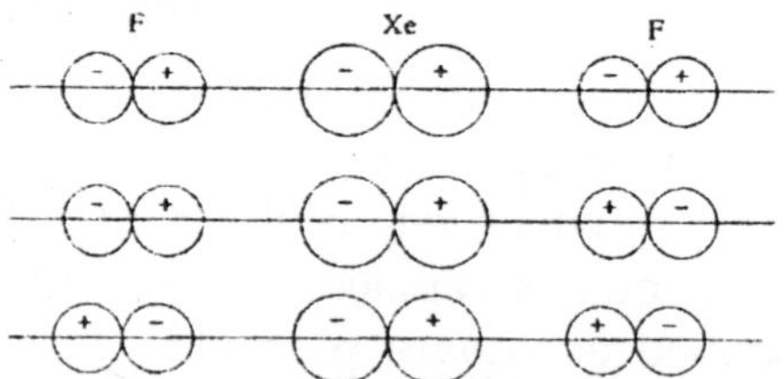

Antibonding (orbitals have wrong symmetry for overlap on both left and right hand sides, indicated by + and – signs)

Non-bonding (the Xe $5p$ orbital has no net contribution to bonding, since the bonding effect on the right hand side is cancelled by the antibonding effect on the left hand side)

Bonding (orbitals on both the left and right hand sides have correct symmetry for overlap)

Possible combinations of atomic orbitals in XeF_2

The original three atomic orbitals contain four electrons (two in the Xe $5p_z$ and one in each of the F $2p_z$). These occupy the molecular orbitals of the lowest energy. Two of these occupy bonding MO and two occupy non-bonding MO, their energies being in the order.

bonding MO < non-bonding MO < antibonding MO.

The bonding may be described as three centre, four electrons σ -bonding. A linear combination of the atoms gives the best overlap of the orbitals.

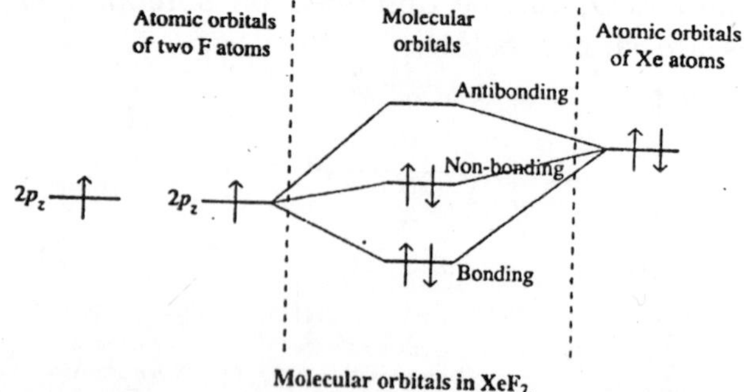

Molecular orbitals in XeF_2.

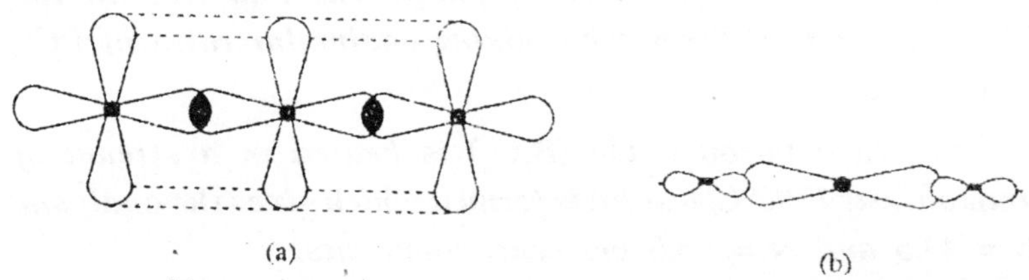

Linear combination of p_z orbitals of xenon and fluorine in XeF_2.

Geometry Around Xenon: Xenon difluoride can act as a fluoride ion donor, and as a result, forms a number of addition compounds with fluoride acceptor such as ASF5 and SbF5. For example,

$$2XeF_2 + SbF_5 \longrightarrow [Xe_2F_3] + \text{ and } [SbF_6]^-$$

The $Xe_2F_3{}^+$ cation has the planar structure as given below:

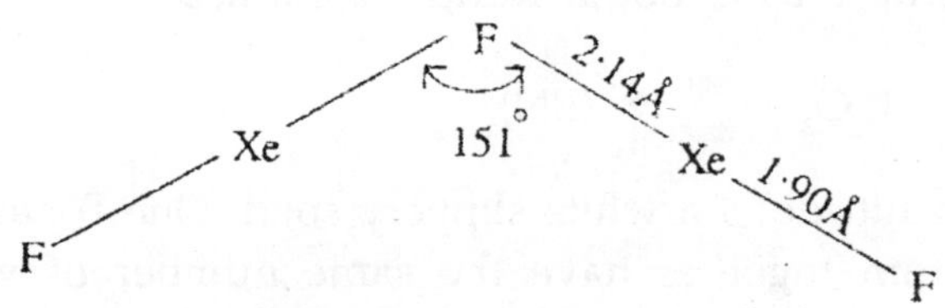

Structure of $Xe_2F_3{}^+$ cation

XeF_4 also form addition compounds with PF_5, AsF_5 and SbF_5.

$$XeF_4 + 3SbF_5 \longrightarrow [XeF_3]^+ + [Sb_3F_{16}]^-$$

The structure of $XeF_3{}^+$ cation can be represnted as follows:

Due to the presence of one unit positive charge on $XeF_2{}^+$ ion, Xe atom can be regarded as having seven electrons (instead of eight)

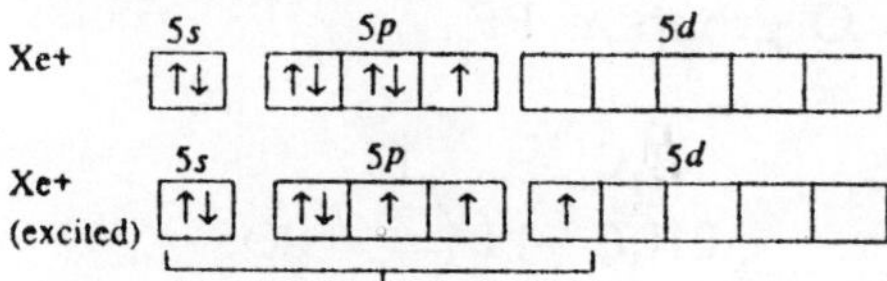

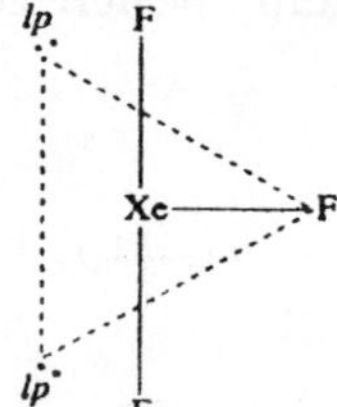

(a) Given the geometry of phophorus and iron in the ioning compound formed by mixing equimolar ratio of PCl_5 and $FeCl_3$.

(b) When boron oxide (B_2O_3) is heated in a stream of ammonia at 1200°C, a solid is formed which gives the analysis: B = 43.5 and N = 56.5 per cent, comments,

(c) Give (i) the products of the reaction between trimethyl-phosphine oxide and sulphur tetrafluoride and (ii) their structures.

(a) $$PCl_5 + FeCl_3 \rightleftharpoons [PCl_4]^+ + [FeCl_4]$$

In $[PCl_4]^+$ the central atom P is sp^3 hybridised and the ionis tetrahedral; the complex $[FeCl_4]^-$ is also tetrahedral.

(b) When boron oxide (B_2O_3) is heated in a stream of ammonia at 1200°C boron nitride is formed.

$$B_2O_3 \xrightarrow[1200°C]{NH_3} (BN)_x$$

Boron nitride is a white slippery solid. One B atom and one N atom together have the same number of valency electrons as two C atoms. Thus boron nitride has almost the same structure as graphite, with sheets made up of hexagonal rings of alternate B and N atoms joined together.

The sheets are staked one on top of theother, giving a layer structure. The formula BN is in completely agreement with the result of analysis, B = 43.5 per cent and N = 56.4.

(c) In the reaction between trimethyl phosphine oxide and sulphur-tetrafluoride, the SF_4 is a selective fluorinaing agent, which converts P = O group to PF_2.

$$2(CH_3)_3\,P = O + SF_4 \longrightarrow 2\,(H_3C)_3PF_2 + SO_2$$

The entries of Column I matched with appropriate entries of Column II.

Column I	*Column II*
1. *Nido compounds*	**A.** *Aluminium alkyl derivative*
2. *Spring reactions*	**B.** *Boranes*
3. *Stock method*	**C.** *Poly phosphates*
4. *Fulminating metal*	**D.** *Cui*
5. *Amphiboles*	**E.** *Cross linked polymers*
6. *Ziegler catalyst*	**F.** *Nitrides of Ag, Au, Hg*
7. *$RSiCl_3$*	**G.** *Hypo*
8. *Electrides*	**H.** *Carboranes*
9. *Rochow Process*	**I.** SмOu⁶"
10. *Grahm's salt*	**J.** *Crown ether complexes*

(1)–(H); (2)–(G); (3)–(B); (4)–(F); (5)–(I); (6)–(A); (7)–(E); (8)–(J); (9)–(D); (10)–(C).

Significance of Silicones

The silicones are a group of organosilicon polymers. The complete hydrolysis of $SiCl_4$ gives SiO_2 which has a stable three dimensional structure. Hydrolysis of alkyl substituted chlorosilane yield long chain polymers called silicones (not the silicon compounds analogous to ketone)

$$R_2SiCl_2 \xrightarrow{+2H_2O} R_2Si(OH)_2$$

$$HO-\underset{R}{\overset{R}{|}}{Si}-OH + HO-\underset{R}{\overset{R}{|}}{Si}-OH \xrightarrow{-H_2O} HO-\underset{R}{\overset{R}{|}}{Si}-O-\underset{R}{\overset{R}{|}}{Si}-OH$$

Since an active OH group is left at each end of chain, polymerisation reaction continues and the length of the chain continue to increase. The starting materials for the manufacture of silicones are *alkyl or aryl substituted chlorosilanes.* Methyl compounds are mainly used, though some phenyl derivatives are used as well. Hydrolysis of dimethyldichlorosilane $(CH_3)_2SiCl_2$ gives rise to straight chain polymers and, as an active OH group is left at each end of the chain, polymerisation continues and the Chain increases in length. $(CH_3)_2SiCl_2$ is therefore a chain building unit. Normally, high polymers are obtained.

$$HO-\underset{CH_3}{\overset{CH_3}{\underset{|}{\overset{|}{Si}}}}-O-\underset{CH_3}{\overset{CH_3}{\underset{|}{\overset{|}{Si}}}}-O-\underset{CH_3}{\overset{CH_3}{\underset{|}{\overset{|}{Si}}}}-O-\underset{CH_3}{\overset{CH_3}{\underset{|}{\overset{|}{Si}}}}-O-\underset{CH_3}{\overset{CH_3}{\underset{|}{\overset{|}{Si}}}}-O-\underset{CH_3}{\overset{CH_3}{\underset{|}{\overset{|}{Si}}}}-O-\underset{CH_3}{\overset{CH_3}{\underset{|}{\overset{|}{Si}}}}-OH$$

Hydrolysis under carefully controlled conditions can produce cyclic structures, with rings containing three, four, five or six Si atoms.

$$\text{cyclo-}[(Me)_2Si-O]_3$$

tris cyclo-dimethylsiloxane

$$\text{cyclo-}[(Me)_2Si-O]_4$$

tetrakis cyclo-dimethylsiloxane

Hydrolysis of trimethylmonochlorosilane $(CH_3)_3SiCl$ yields $(CH_3)_3SiOH$ trimethylisilanol as a volatile liquid, which can condense, giving hexamethyldisiloxane. Since this compound has no OH groups, it cannot polymerise any further.

$$CH_3-\underset{CH_3}{\overset{CH_3}{\underset{|}{\overset{|}{Si}}}}-OH + HO-\underset{CH_3}{\overset{CH_3}{\underset{|}{\overset{|}{Si}}}}-CH_3 \rightarrow CH_3-\underset{CH_3}{\overset{CH_3}{\underset{|}{\overset{|}{Si}}}}-O-\underset{CH_3}{\overset{CH_3}{\underset{|}{\overset{|}{Si}}}}-CH_3$$

Hexamethyldisiloxane

If some $(CH_3)_3$ SiCl is mixed with $(CH_3)_2$ $SiCl_2$ and hydrolysed, the $(CH_3)_3SiCl$ will block the end of the straight chain produced by $(CH_3)_2SiCl_2$, Since there is no longer a functional OH group at this end of the chain, it cannot grow any more at this end. Eventually the other end will be blocked in a similar way. Thus $(CH_3)_3SiCl$ is a chain stopping unit, and the ratio of $(CH_3)_3SiCl$ and $(CH_3)_2SiCl_2$ in the starting mixture will determine the average chain size.

The hydrolysis of methyl trichlorosilane $RSiCl_3$ gives a very complex cross-linked polymer.

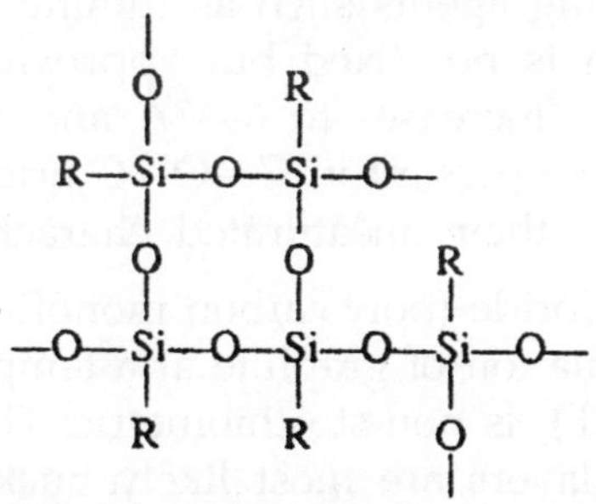

```
           |
           O       R
           |       |
       R—Si—O—Si—O—
           |       |
           O       O       R
           |       |       |
     —O—Si—O—Si—O—Si—O—
           |       |       |
           R       R       O
                           |
```

It may be noted that unlike carbon which cannot hold more than one OH group, the silicon atom in the compound $RSi(OH)_3$ can hold three OH groups. It is this property of silicon that makes the formation of organosilicon polymers possible. The strength and inertness of silicones are related to two factors.

1. Their stable silica-like skeleton of Si—O—Si—O—Si. The Si—O bond energy is very high (502 kJ mol^{-1}).
2. The high strength of the Si—C bond.

Silicones were originally developed as electrical insulators, because they are more stable to heat than are organic polymers, and if they do break down they do not produce conducting materials as carbon does. They are resistant to heat, oxidation and most chemicals. They are strongly water repellent, are good electrical insulators, and have non-stick properties and anti-foaming properties. Their strength and inertness are related to two factors.

Graphite Compounds

A very loose, layered structure of graphite makes it possible for many molecules or ions to penetrate between the layers, forming interstitial or lamellar compounds.

These are of two basic types : those in which the graphite, which has good conducting properties, becomes non-conducting and those in which high electrical conductivity remains and is enhanced. Only two substances of first types are known, namely graphite oxide and graphite fluoride. Graphite oxide is obtained by treating graphite with strong aqueous oxidising agents such as fuming HNO_3 or $KMnO_4$. Its composition is not fixed but approximates to C_2O, the layer separation increases to 6—7Å and it is believed that oxygen atoms are present in C—O—C bridges ; the graphite layers thus lose their unsaturated character and buckle.

Graphite fluoride (poly carbon monofluoride) is obtained by direct fluorination of graphite at a temperature of ~ 600°. The product $(CF)_n$ is non-stoichiometric. The layer spacing is ~ 8Å and the layers are most likely buckled.

In electrically conducting lamellar compounds, various atoms, molecules or ions are inserted between the carbon sheets. A large number of compounds are formed spontaneously when graphite and the reactants are brought into contact. Alkali metals, halogens, $FeCl_3$, $UC1_4$, FeS_2 and MoO_3 form lamellar compounds spontaneously. The manner in which the invading reactant species increase the conductivity of graphite is not definite, but apparently they do so by either adding electrons to or remaining electrons from the conduction level of graphite.

Spring Reaction: It is used for the preparation of sodium thiosulphate. It consists in treating a mixture of sodium sulphide and sodium sulphite with calculated quantity of iodine.

$$Na_2S + Na_2SO_3 + I_2 \longrightarrow Na_2S_2O_3 + 2NaI$$

Lonic Mobility: Lonic mobility or conductivity measurements in aqueous solution give results in the order

$$Cs^+ > Rb^+ > K^+ > Na^+ > Li^+$$

The reason for this anomaly is that the ions are hydrated in solutions. Since Li^+ is very small, it is heavily hydrated. This makes the radius of hydrated ion large, and hence it moves very slowly. Cs^+ is least hydrated and radius of hydrated Cs^+ ion is smaller.

Sulphur is reduced by a solution of sodium in liquid NH_3 but not by a solution of sodium in water. This is because alkali metals are stronger reducing agents than hydrogen and so will react with water and liberate hydrogen. The metals can exist for some time in liquid ammonia.

$$S + Na/NH_3 \longrightarrow Na_2S$$

S is reduced from oxidation state O to –II.

In the crown ether complex dibenzo-18-crown-6 there are two benzene rings and 18 atoms makes up a crown-shaped ring and six of the ring atoms are oxygen. These six oxygen atoms may complex with a metal ion with larger metal ions of group I. Similarly benzo-12-crown-4 has a ring of 12 atoms, four of which are oxygen.

The structure of these compounds are as:

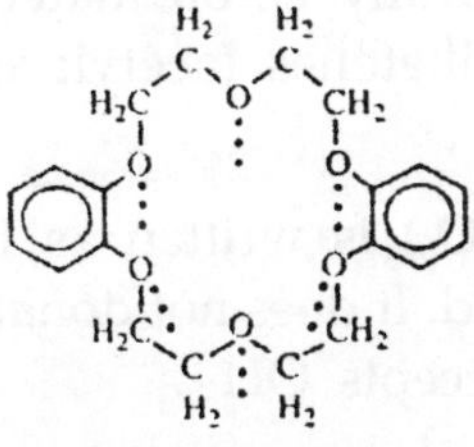

dibenzo-18-crown-6

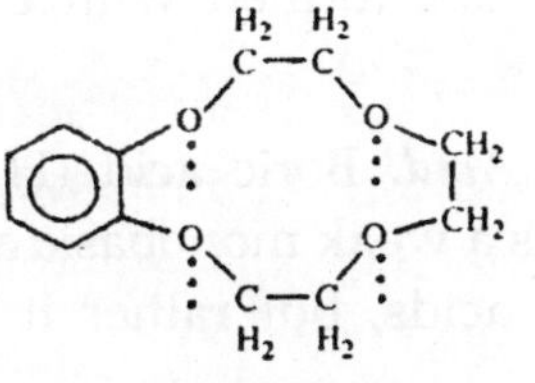

benzo-12-crown-4

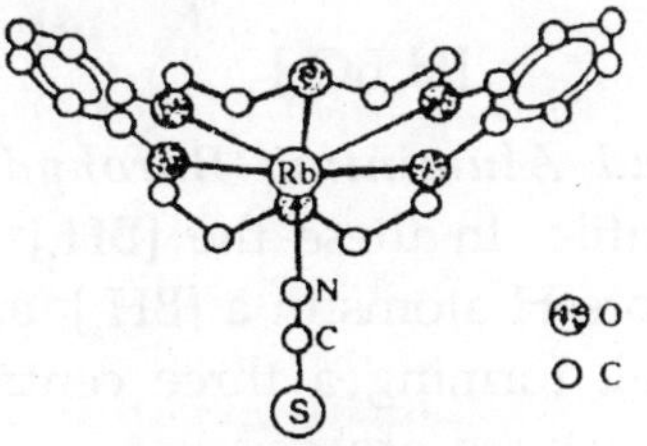

RbCNS(dibenzo-18-crown-6) complex

Structure of Basic Beryllium Acetate: Basic beryllium acetate $[Be_4O(CH_3COO)_6]$ is formed if $Be(OH)_2$ is evaporated with acetic acid. The structure comprises a central oxygen atom surrounded by four beryllium atoms located at the corners of a tetrahedron, with six acetate groups (Ac) arranged along the six edges of the tetrahedron. Similar structure is there for a series of stable covalent molecules of formula $[Be_4O(R_6)]$ where R may be NO_3^-, $HCOO^-$, CH_3COO^-, $C_2H_5COO^-$ $C_6H_5COO^-$, etc.

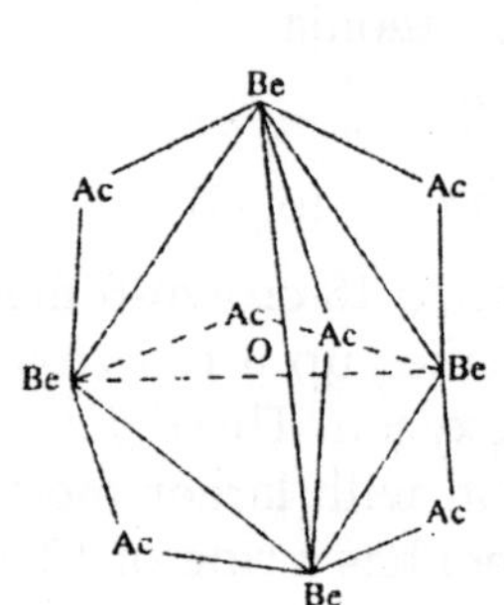

(a) Basic beryllium acetate $[Be_4O(CH_3COO)_6]$

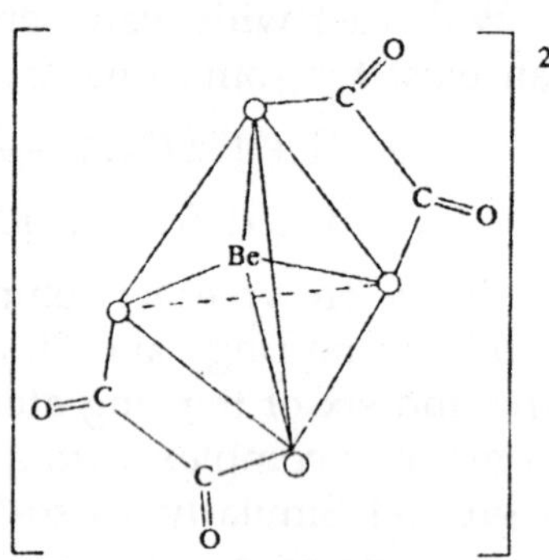

(b) Beryllium oxalate $[Be\,(OX)_2]^2$

Structure of Beryllium Oxalate: In beryllium oxalate $[Be(OX)_2]^{2-}$, Be^{2+} ion is tetrahedrally co-ordinated. Similar structures are formed with b-diketones (acetylacetone and catechol).

Boric Acid: Boric acid (H_3BO_3) is written as $B(OH)_3$. ft behaves as a weak monobasic acid. It does not donate protons like most acids, but rather it accepts OH~.

$$B(OH)_3 + 2H_2O \rightleftharpoons [B(OH)_4]^-$$

or, $[H_3BO_3]$

Beryllium and Aluminium Borohydrides: These are covalent and volatile. In these the $[BH_4]^-$ groups acts as a ligand. One or more H atoms in a $[BH_4]^-$ act as a bridge and bond to the metal, forming a three centre bond with two electrons. The structures are as

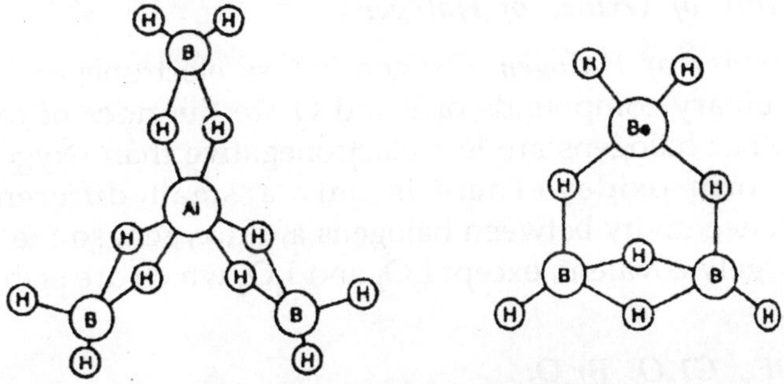

Structure of $Al(BH_4)_3$ and $Be(BH_4)_2$

Boron nitride is a white slippery solid. One B atom and one N atom together have the same number of valency electrons as two C atoms. Thus boron nitride has almost the same structure as graphite, with sheets made up of hexagonal rings of alternate B and N atoms joined together.

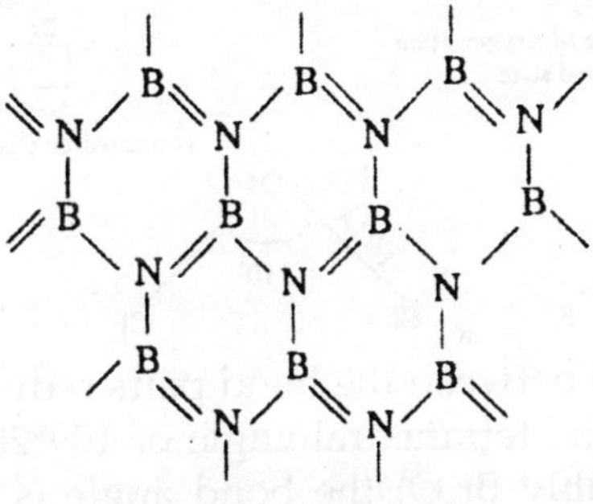

Boron nitride

Borazine ($B_3N_3H_6$) is sometimes called 'inorganic benzene' because its structure shows some formal similarity with benzene, with delocalised electrons and aromatic character.

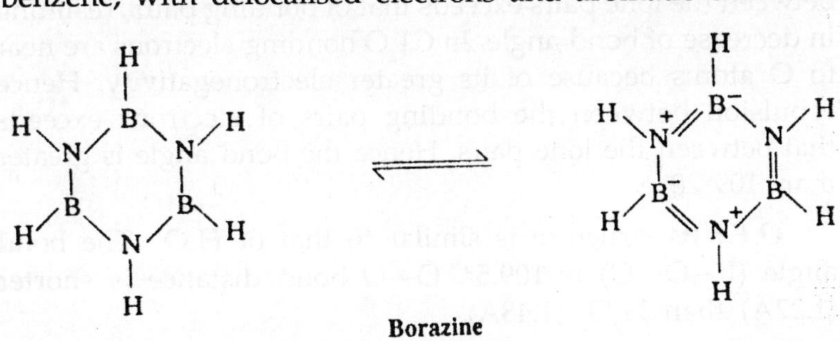

Borazine

Structure of Oxides of Halogens

Oxides of Halogen: Oxygen is less electronegative than F, so binary compounds of F and O are fluorides of oxygen. The other halogens are less electronegative than oxygen and thus form oxides. There is only a small difference in electronegativity between halogens and oxygen, so the bonds are largely covalent, except I_2O_4 and I_4O_9 which are stable and ionic.

OF_2, $C1_2O$, Br_2O:

$$2F_2 + 2OH \longrightarrow 2F^- + F_2O + H_2O$$

$$2HgO + 2Cl_2 \longrightarrow HgCl_2 \cdot HgO + C1_2O$$

The structures of OF_2, $C1_2O$ and Br_2O are all related to a tetrahedron with two positions occupied by alone pair of electrons.

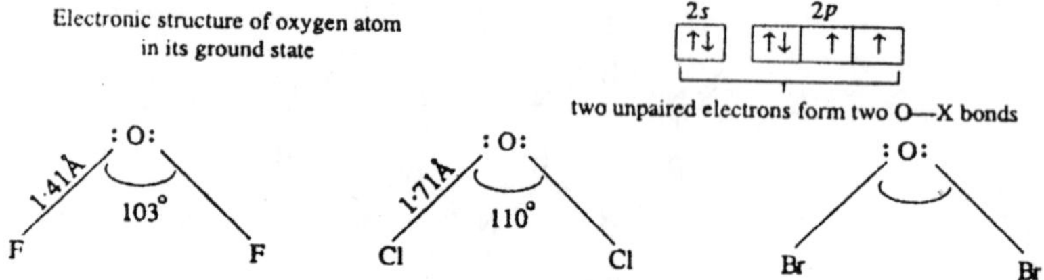

Repulsion between the bond pairs reduces the bond angle in F_2O from the tetrahedral angle of 109°28' to 103°. In $C1_2O$ (and presumably Br_2O) the bond angle is increased because of steric crowding of larger halogen atoms.

In other way in OF_2, the bonding electrons are nearer of F atom because of higher electronegativity, so the repulsion between the lone pairs exceeds that of bonding pairs, resulting in decrease of bond angle. In $C1_2O$ bonding electrons are near to O atoms because of its greater electronegativity. Hence repulsion between the bonding pairs of electrons exceeds that between the lone pairs. Hence the bond angle is greater than 109°28'.

O_2F_2: Its structure is similar to that of H_2O_2. The bond angle (F—O—O) is 109.5°. O—O bond distance is shorter (1.22Å) than H_2O_2 (1.48Å)

F
|1·58Å
122Å O
O
F

***C1O₂*:** It is formed by the following reaction:

$$3C1O_3^- + 2H^+ \longrightarrow 2C1O_2 + ClO_4^- + H_2O$$

It is paramagnetic and contains an odd number of electrons and is, therefore, highly reactive. $C1O_2$ molecule is analogous to that of NO_2 in that it contains a three electron bond and has an angular shape but $C1O_2$ does not dimerise probably because the odd electron is delocalised. The bond lengths are both 1.47Å and are shorter than for single bond and resonance is thought to occur between two structures. The Cl—O—Cl bond angle is 118°.

(Cl—O π bond is *pπ-dπ* bond)

Cl_2O_6—The oxide is diamagnetic and possible structures are :

***$C1_2O_7$*:** The molecule is polar and electron diffraction analysis reveals that two $C1O_4$ tetrahedra are linked through oxgen atom in its structure.

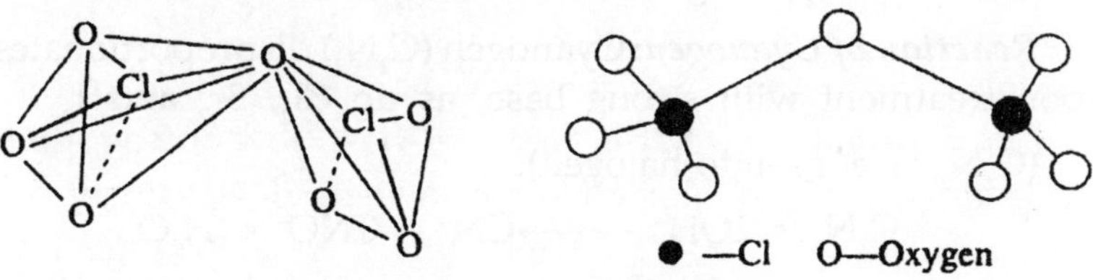

***I_2O_5*:** I_2O_5 solid is a three dimensional network with strong intermolecular I...O interactions linking molecules together.

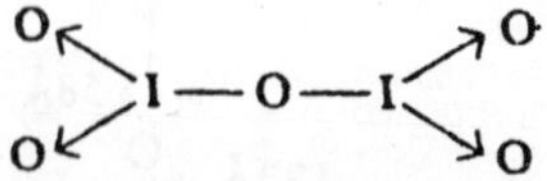

Bond Angle in $C1_2O$ is Less than $C1O_2$: This is because the repulsion for the bonded pairs is less in $C1O_2$ than that in $C1_2O$ due to non-bonding electrons on the central atom in $C1O_2$. The shorter bond length in $C1O_2$ results from resonance, with the unpaired electron on the chlorine or the oxygen atom.

Preparation of Compounds:

(a) $3C1_2 + 6NaOH \longrightarrow 5NaCl + NaC1O_3 + 3H_2O$

$4NaClO_3 \xrightarrow{\text{Heat}} NaCl + 3NaC1O_4$

$4NaClO_4 + H_2SO_4 \longrightarrow NaHSO_4 + HC1O_4$

(b) $I_2 + 6OH^- \longrightarrow 5I^- + IO_3^- + 3H_2O$

$IO_3^- + H^+ \longrightarrow HIO_3$

$2HIO_3 \xrightarrow{\text{Heat}} I_2O_5 + H_2O$

(c) $2C1_2^- + 2HgO \longrightarrow HgCl_2.HgO + C1_2O$

(d) $C1_2 + 6OH^- \longrightarrow 5Cl^- + ClO_3^- + 3H_2O$

$3C1O_3 + 2H^+ \longrightarrow 2ClO_2 + ClO_4^- + H_2O$

(e) $3Br_2 + 6KOH \longrightarrow 5KBr + KBrO_3$

(f) $2F_2 + 2OH^- \longrightarrow 2F^- + F_2O + H_2O$

(g) $\frac{1}{2}Br_2 + O_3 \xrightarrow{0°C} BrO_3$

(h) $2Br_2 + HgO \longrightarrow Br_2O + HgO$

Reaction of Cyanogen: Cyanogen (C_2N_2) disproportionates upon treatment with strong base, as do Cl_2, Br_2 and I_2

(C_2N_2 is a pseudo-halogen).

$$C_2N_2 + 2OH^- \longrightarrow CN^- + CNO^- + H_2O$$

Structure of Silicates

Silicates are the compounds in which the anions present are either discrete SiO_4^- tetrahedra or a number of such units joined together through corners. These discrete tetrahedra or a number of tetrahedra linked together are present as the anions in the silicates. However, when they link together, they do so by corners and never by edges. Thus the anions present in the silicates may be SiO_4^{4-} (for discrete tetrahedra), $Si_2O_7^{6-}$ (for two tetrahedra linked through one corner) and so on.

Classification of Silicates: On the basis of corners (0, 1, 2, 3 or 4) of the SiO_4 tetrahedra shared with other neo-silicate.

Orthosilicates [SiO$_4$]$^{4-}$: They contain discrete $(SiO_4)^{4-}$ tetrahedra, *i.e.,* they share no corners (Fig.)

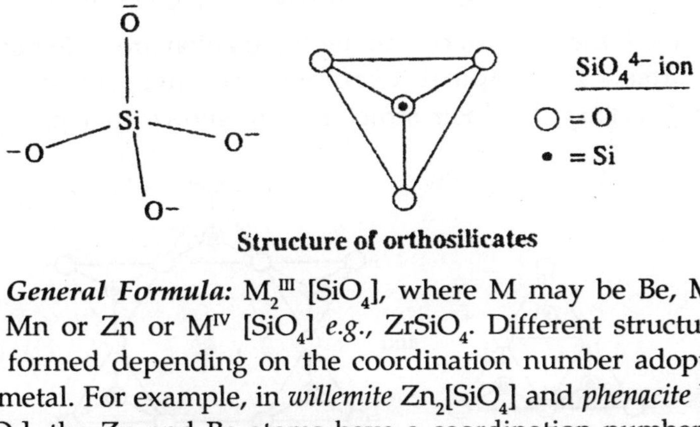

Structure of orthosilicates

General Formula: M_2^{III} [SiO_4], where M may be Be, Mg, Fe, Mn or Zn or M^{IV} [SiO_4] *e.g.,* $ZrSiO_4$. Different structures are formed depending on the coordination number adopted by metal. For example, in *willemite* $Zn_2[SiO_4]$ and *phenacite* Be_2 [SiO_4], the Zn and Be atoms have a coordination number of 4 and occupy tetrahedral holes.

In *forsterite* $Mg_2[SiO_4]$, the Mg has a coordination number of 6 and occupies octahedral holes. When octahedral sites are occupied, it is quite common to get isomorphous replacement of one divalentmetal ion by another of similar size, without changing the structure Zircon $ZrSiO_4$ has a coordination number *S*. The structure is not close packed.

Pyrosilicates (Si$_2$O$_7$)$^{6-}$ - soro-silicates, Disilicates: Two tetrahedral units are joined by sharing the O at one corner, thus giving $(Si_2O_7)^{6-}$. The structure possessed by them are called island silicates structure (Fig)

Si $_2$O $_7$ $^{6-}$ ion

Structure of pyrosilicates $Si_2O_7{}^{6-}$

Example: ***Therteveitite*** ***$Sc_2(Si_2O_7)$*****,** in this Sc^{3+} ions are octahedrally coordinated. *Hemimorphite* $Zn_4(OH)_2\ (Si_2O_7).\ H_2O$.

Cyclic or Ring Silicates: If two oxygen atoms per tetrahedron are shared to form closed rings such that the structures with the general formula $\left(SiO_3^{2-}\right)_n$ or $\left(SiO_3\right)_n^{2n-}$, are obtained, the silicates containing these anion are called cyclic silicates. The typical examples of such anions are $Si_3O_9^{6-}$ and $Si_6O_{18}^{-12}$. Their structures are shown in Fig.

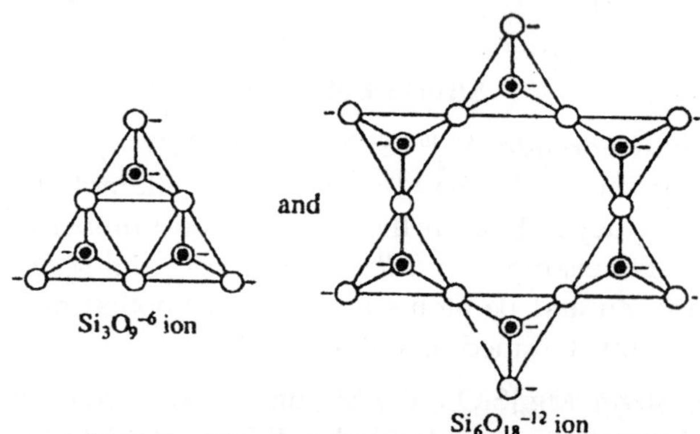

Structure of cyclic silicates $Si_3O_9{}^{-6}$ and $Si_6O_{18}{}^{-12}$

Examples: *Wollartnite* $Ca_3\ (Si_3O_9)$, *Benitoite* BaTi $[Si_3O_9]$, *Beryl* $Be_3Al_2\ [Si_6O_{18}]$.

Chain Silicates, Linearpolyanion, Metasilicates: Simple chain silicate or pyroxenes are formed by the sharing of the O atoms on two corners of each tertahedron with other tetrahedra. This gives the formula $\left(SiO_3\right)_n^{2n-}$, Fig.

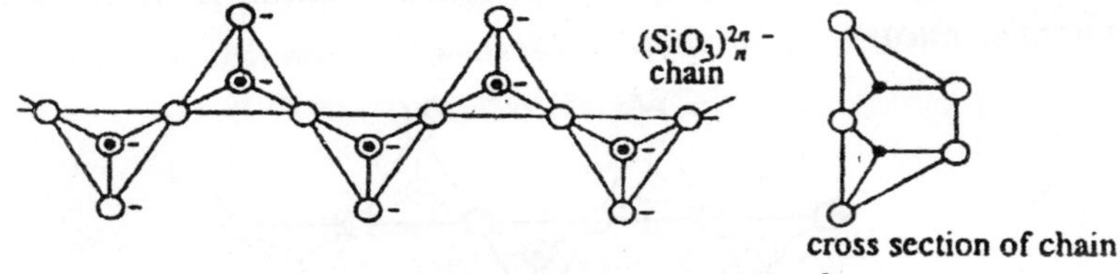

Structure of pyroxenes $(SiO_3)_n^{2n-}$

Examples: *Spodumene* LiAl $[(SiO_3)_2]$, *Diopsite* CaMg$[(SiO_3)_2]$, *Wollastonite* $Ca_3[(SiO_3)_3]$

Double chains can be formed when two simple chains are joined together by shared oxygen. These minerals are called *amphiboles*. There are several ways of forming double chains giving formulae like

$(Si_4O_{11})_n^{6n-}$ and others. $(Si_4O_{11})_n^{6n-}$, is shown in Fig.

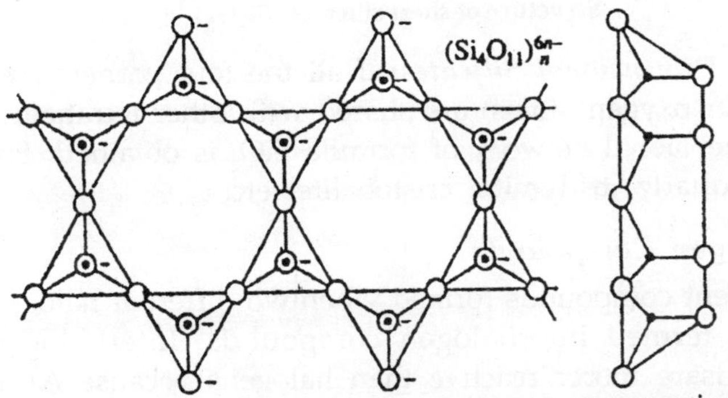

Structure of amphiboles $(Si_4O_{11})_n^{6n-}$

A typical amphibole is termolite $Ca_2Mg_5(Si_4O_{11})2\ (OH)_2$. Asbestos minerals are all amphiboles.

Sheet Silicates: The sharing of three corners (*i.e.,* three oxygen of each tetrahedron results in an infinite two dimensional sheet structure of the formula $(Si_2O_5)_n^{2n-}$ Fig. There are strong bonds within the Si—O sheet, but much weaker forces hold each sheet to the next one. Thus these minerals tend.to cleave into thin sheets. These include *clay minerals, mica.*

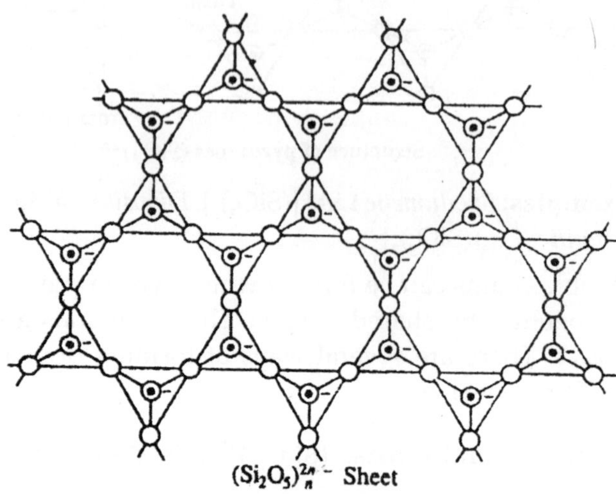

$(Si_2O_5)_n^{2n-}$ Sheet

Structure of sheet silicates $(Si_2O_5)_n^{2n-}$

Three Dimensional Silicates: If all the four corners (*i.e.,* all the four oxygen atoms) are shared with other tetrahedra, three dimensional network of formula SiO_2 is obtained. For example quartz, tridymite, cristobalite, etc.

Interhalogen Compounds

Covalent compounds formed whentwo different halogen react are termed interhalogen compounds. Interhalogen compoundsare morer reactive than halogens because A—B bond is weaker than B—B.

Significance of Carboranes

The carboranes are mixed hydrides of carbon and boron having both carbon and boron atoms in electron deficient skeletal framework.

The polyhedral carboranes may be considered as formally derived from the $B_nH_n^{2-}$ ions on the basis that the CH group is isoelectronic and isostructural with, and may thus replace, the BH^- group.

Geometrically carboranes are classified into two types:

(i) Those in which the boron atom framework closes in on itself to form a polyhedron. These are *closo* (cage) compounds. These have the general formula C_2B_{n-2} H_n $(n = 5 - 12)$.

(ii) Those which have the open cage structures, derived formally from one or other of several boranes and containing from one to four carbon atoms in the skeleton. These are *nido* (nest) compounds.

The boranes most commonly used in making the smaller carboranes are B_4H_{10}, B_5H_9 and B_5H_{11}. For example, B_5H_9 and C_2H_2 react in gas phase at 215°C to give mainly the *nido* carborane 2, 3 -$C_2B_4H_8$. The same reactants at 450° or in an electric discharge give the **closo** carborane 1,5- $C_2B_3H_5$ 1, 6-$C_2B_4H_6$ and 2, 4-$C_2B_5H_7$. The nido carborane 2, 3-$C_2B_4H_8$ is converted to the coloso-carboranes $C_2B_3H_5$, $C_2B_4H_6$ and $C_2B_5H_7$ on pyrolysis orultraviolet irradiations.

The closo-carboranes of the $C_nB_{n-2}H_n$ series are isoelectronic with corresponding $[B_nH_n]^{2-}$ ions and have the same closed polyhedral structures with one hydrogen atom bonded to each carbon and boron.

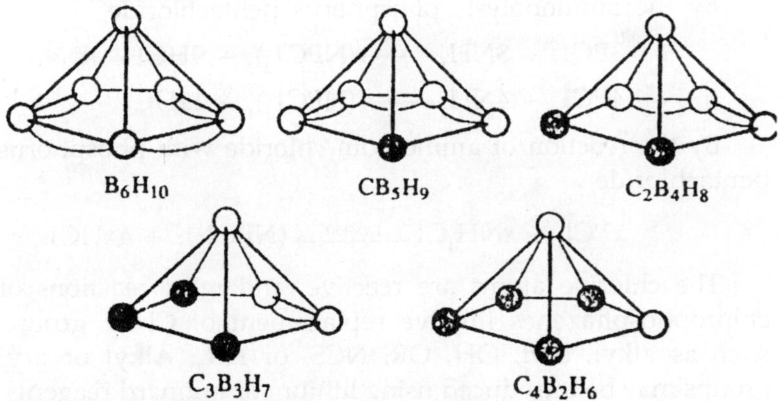

The *nido* -carboranes formally related to B_6H_{10}. All have eight pairs of electrons bonding the six cage atoms together. Hydrogen bridges are represented by curved lines.

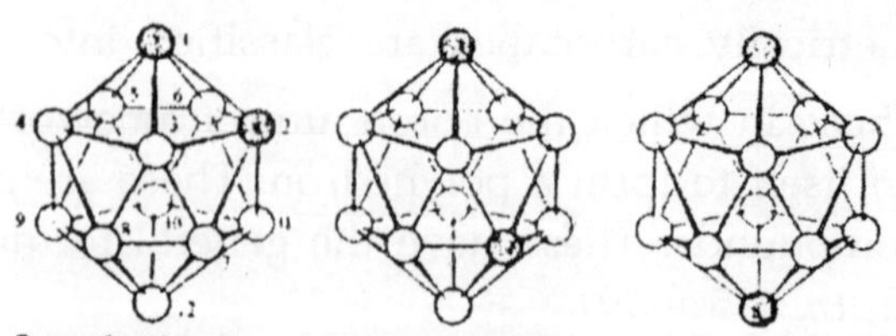

1, 2-(*ortho*) isomer 1, 7-(*meta*) isomer 1, 12- (*para*) isomer

The three dicarba-*closo*-dodecaboranes, $C_2B_{10}H_{12}$

An account of phosphazenes with their structural aspects:

Phosphazenes or Phosphonitrilic Compounds: Phosphazenes are a group of compounds represented by the general formula $(NPX_2)_n$ where X = F, Cl, Br, SCN, CH_3, C_6H_5, etc. All these compounds are polymeric, in these compounds P atom is in oxidation state (+V) and N is in the (+ III) state. The compounds are formally unsaturated.

Phosphonitrilic chlorides are important phosphazenes with the general formula $(NPC1_2)$J where *x* ranges from 3 to 7 and upwards. The monomer (x = 1) and dimer (x = 2) are not known. These compounds were originally called phosphonitrilic halides, but are now named systematically poly (chlorophosphazenes).

These can be prepared as follows:

By the ammonolysis phosphorus pentachloride

$$3PC1_5 + 3NH_3 \longrightarrow (NPC1_2)_3 + 9HC1$$

$$4PC1_5 + 4NH_3 \longrightarrow (NPC1_2)_4 + 12HC1$$

By the reaction of ammonium chloride with phosphorus pentachloride

$$xPCl_5 + xNH_4C1 \xrightarrow{120\text{–}150^\circ} (NPC1_2)_X + 4xHC1$$

The chlorine atoms are reactive, and most reactions of chlorophosphazenes involve replacement of Cl by groups such as alkyl, aryl, OH, OR, NCS, or NR_2. Alkyl or aryl groups may be introduced using lithium or grignard reagents.

$$(NPCl_2)_3 + 6CH_3MgI \longrightarrow [NP(CH_3)_2]_3 + 3MgCl_2 + 3MgI_2$$

$$(NPCl_2)_3 + 6C_6H_5Li \longrightarrow [NP(C_6H_5)_2]_3 + 6LiCl$$

$$(NPCl_2)_3 + 6NaOR \longrightarrow [NP(OR)_2]_3 + 6NaCl$$

On hydrolysis chlorine atoms of chlorophosphazenes can replaced by —OH groups.

Structure: X-ray examination reveals that the trimer and tetramer chlorophosphazenes are having the six and eight membered rings and composed of alternate nitrogen and phosphorus atoms. Thus these compounds may be regarded as the phosphorus-nitrogen analogues of benzene and cyclo-octatetraene

$(NPC1_2)_3$ has almost planar ring. There is an approximately tetrahedral distribution of valencies round phosphorus corresponding with sp^3 hybridisation with the result that the chlorine atoms in each pair lie on opposite side of the rings. All the N—P bonds in the ring are also of equal length, *i.e.*, 1.59 ± 0.02Å.

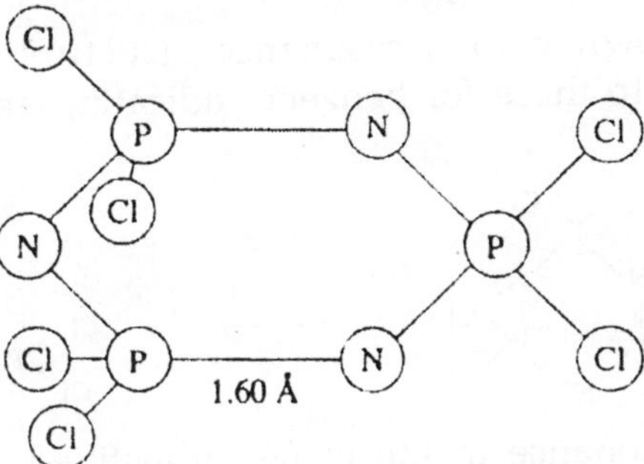

The Structure of $N_3P_3Cl_6$ trimer.

In the above structure it is evident that two chlorine atoms are attached to each phosphorus atom. This is confirmed by two chemical experiments such as:

(i) *Benzene reacts with* $N_3P_3C_{16}$ *in the presence of* $AICl_3$ thus:

$$\underset{\text{(II)}}{N_3P_3Cl_6} + 2C_6H_6 \longrightarrow \underset{\text{(III)}}{N_3P_3Cl_4(C_6H_5)_2} + 2HCl$$

The product (III) is hydrolysed by water to $(C_6H_5)_2$ PO.OH. Hence both the chlorine atoms originally have been attached to the same phosphorus atom.

(ii) *Phenyl magnesium bromide at* 115° *also reacts with* $N_3P_3C1_6$ *to produce diphenyl derivatives.*

When molecular orbital theory is applied to the structure of $N_3P_3C1_6$, it is observed that there are normal localised *a* bonds through the ring formed by overlap of sp^3 hybrid orbitals of phosphorus with sp^2 orbital of nitrogen. In addition, each introgen atom has an electron in a p_z orbital and each phosphorus atom in a *d*-orbital. These orbitals combine to give delocalised $p\pi$ - $d\pi$ orbitals which extend over the whole ring structure.

The presence of negative chlorine atoms on phosphorus makes the diffuse phosphorus *d*-orbitals more compact and favours their overlap.

As shown below, resonance structures can be drawn analogous to those for benzene indicating aromaticity in the ring

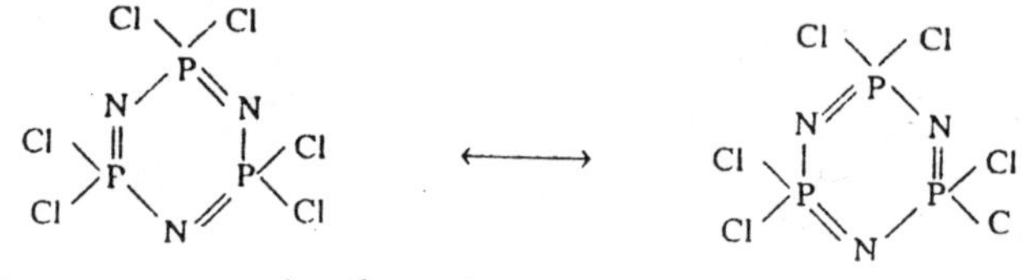

The resonance in the trimer is justified by the fact that all the N—P distances are the same (1-67 Å) which are less than that expected for a P—N single bonds (1.78°A) as in sodium phosphoramidate.

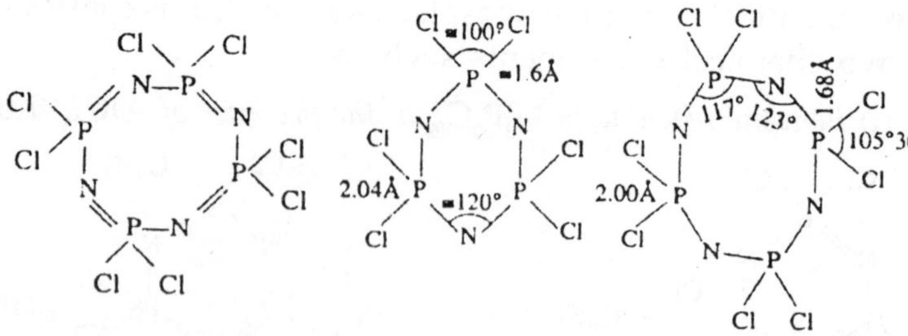

The Structure of trimer and tetramer of phosphonitrilic chlorides.

The tetrameric chloride $(NPC1_2)_4$ has a remarkable puckered ring structure of alternate phosphorus and nitrogen atoms, with two chlorines on each phosphorus atom.

$[NPC1_2]_3$ is flat and $[NPC1_2]_4$ exists in the chair and boat conformation.

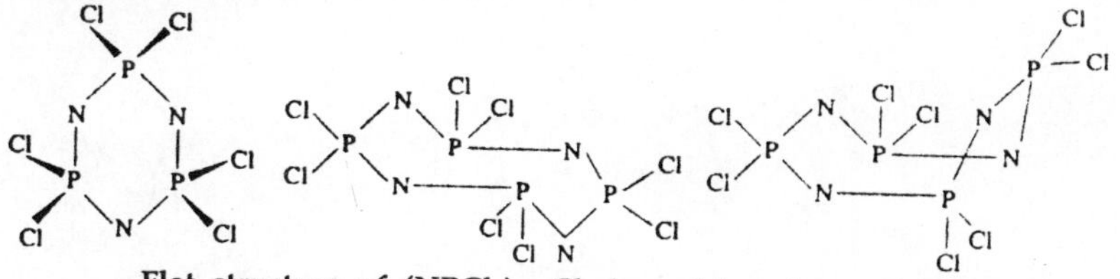

Flat structure of $(NPCl_2)_3$; Chair and bot form of $(NPCl_2)_4$

The reaction completed and given the structure of the main product:

$$[PNCl_2]_3 + \text{excess } (CH_3)_2\ NH \rightarrow$$

$$[PNC1_2]_3 + \text{excess } (CH_3)_2\ NH \rightarrow [P_3N_3\{(CH_3)_2N\}_6]$$

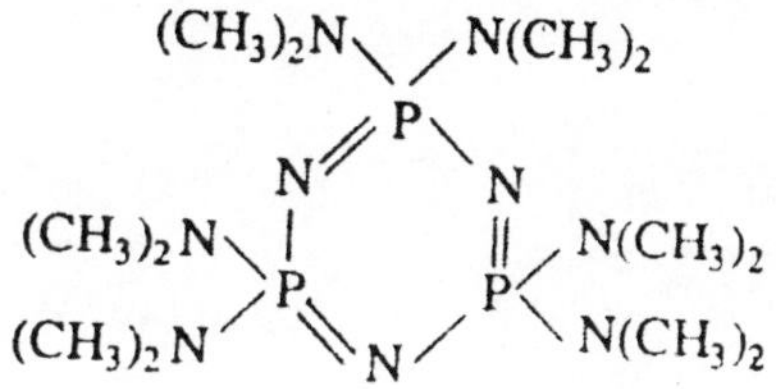

Bibliography

Alan, J. Rocke: *Chemical Atomism in the Nineteenth Century*, Ohio State University Press, Columbus, 1982.

Allibone, T. E.: *Rutherford: The Father of Nuclear Energy*, Manchester University Press, Manchester, 1973.

Archibald, Clow and Nan L. C.: *The Chemical Revolution*, Gordon & Breach Science Publishers, London, 1992.

Arnold, Thackray: *John Dalton: Critical Assessments of His Life and Science*, Harvard University Press, Cambridge, 1972.

Arthur, Donovan: *Antoine Lavoisier: Science, Administration and Revolution*, Cambridge University Press, New York, 1996.

Bard, A. J. and Mirkin, M. V.: *Scanning Electrochemical Microscopy*, John Wiley & Sons, New York, 2001.

Bard, A. J. and Zhan W.: *Chemically Imaging Living cells by Scanning Electrochemical Microscopy*, Biosensors & Bioelectronics, 2006.

Benjamin, Loeb S.: *The Atomic Energy Commission under Nixon: Adjusting to Troubled Times*, St. Martin's Press, New York, 1993.

Brandt, E. N.: *Growth Company: Dow Chemical's First Century*, Michigan State University Press, East Lansing, 1997.

Brenda, Buchanan: *Gunpowder: The History of an International Technology*, Bath University Press, United Kingdom, 1996.

Carol, Moberg L.: *The Beckman Symposium on Biomedical Instrumentation*, Beckman Instruments, Fullerton, 1986.

Chang, Raymond: *Chemistry*, James M. Smith, Boston, 1998.

Charles, C. C.: *Alcoa: An American Enterprise*, Rinehart, New York, 1952.

David, E. Newton: *Linus Pauling: Scientist and Advocate*, Facts on File, New York, 1994.

David, Knight: *Humphry Davy: Science and Power*, Cambridge University Press, New York, 1996.

David, Wilson: *Rutherford, Simple Genius*, MIT Press, Cambridge, 1983.

Deborah, Crawford: *Lise Meitner: Atomic Pioneer*, Crown Publishers, New York, 1969.

Dietrich, Stoltzenberg: *Fritz Haber: Chemist, Nobel Laureate, German, Jew*, Chemical Heritage Foundation, Philadelphia, 2004

Edward, Acheson G.: *A Pathfinder*, Acheson Industries, Port Huron, 1965.

Elisabeth, Crawford: *Arrhenius: From Ionic Theory to the Greenhouse Effect*, Science History Publications, Canton, 1996.

Elizabeth, C. Patterson: *John Dalton and the Atomic Theory*, Doubleday, Garden City, 1970.

Engstrom, R.C. and Pharr C.M.: *Scanning Electrochemical Microscopy*, Analytical Chemistry, 1989.

Ernst, Baumler: *Paul Ehrlich: Scientist for Life*, Holmes & Meier, New York, 1984.

Evan, Melhado M. and Tore Frangsmyr: *Enlightenment Science in the Romantic Era*, Cambridge University Press, New York, 1992.

Evan, Melhado M.: *Jacob Berzelius: The Emergence of His Chemical System*, University of Wisconsin Press, Madison, 1981.

Fan F. R. and Mirkin M.V.: *Scanning Electrochemical Microscopy in Electroanalytical Chemistry*, Marcel Dekker, New York, 1994.

Francis, Crick H.: *What Mad Pursuit: A Personal View of Scientific Discovery*, Basic Books, New York, 1988.

Frederic, Holmes L.: *Lavoisier and the Chemistry of Life: An Exploration of Scientific Discovery*, University of Wisconsin Press, Madison, 1985.

G. M. Caroe: *William Henry Bragg, 1862–1942: Man and Scientist*, Cambridge University Press, New York, 1978.

George, David Smith: *From Monopoly to Competition: The Transformations of Alcoa, 1888–1986*, Cambridge University Press, Cambridge, 1988.

Glenn, T. Seaborg: *A Chemist in the White House: From the Manhattan Project to the End of the Cold War*, American Chemical Society, Washington, 1996.

Goodstein, Judith R.: *Millikan's School: A History of the California Institute of Technology*, Norton, New York, 1991.

Haber, L. F.: *The Poisonous Cloud: Chemical Warfare in the First World War*, Oxford University Press, New York, 1986.

Hager, Thomas: *Force of Nature: The Life of Linus Pauling*, Simon and Schuster, New York, 1995.

Harold, Hartley: *Sir Humphry Davy*, Nelson, London, 1966.

Harrison, Stephens: *Golden Past, Golden Future: The First Fifty Years of Beckman Instruments*, Claremont University Press, California, 1985.

Heisenberg, Werner: *On the Perceptual Content of Quantum Theoretical Kinematics and Mechanics*, Princeton University Press, Princeton, 1983.

Heitler, Walter and London Fritz: *Interaction of Neutral Atoms and Homopolar Binding According to the Quantum Mechanics*, 1927.

Henry, Guerlac: *Lavoisier—The Crucial Year: The Background and Origin of His First Experiments on Combustion in 1772*, Cornell University Press, Ithaca, 1961.

Jammer, Max.: *The Conceptual Development of Quantum Mechanics*, McGraw-Hill, New York, 1966.

Jean-Pierre, Poirier: *Lavoisier: Chemist, Biologist, Economist*, University of Pennsylvania Press, Philadelphia, 1996.

John, Stock T. and Mary Virginia Orna: *Electrochemistry Past and Present*, American Chemical Society, Washington, 1989.

John, W. S.: *Physical Chemistry from Ostwald to Pauling: The Making of a Science in America*, Princeton University Press, Princeton, 1990.

Joseph, Priestley: *A Scientific Autobiography of Joseph Priestley, 1733–1804, Selected Scientific Correspondence*, MIT Press, Cambridge, 1966.

Junius, David Edwards: *Immortal Woodshed: The Story of the Inventor Who Brought Aluminum to America*, Dodd, New York, 1955.

Kamb, Barclay: *Linus Pauling, Selected Scientific Papers*, World Scientific, New Jersey, 2001.

Kenne, Fant: *Alfred Nobel: A Biography*, Arcade, New York, 1993.

Laforge, F. O. and Mirkin M. V.: *Scanning Electrochemical Microscopy in the 21st Century*, Physical Chemistry Chemical Physics, 2007.

Lewis, Gilbert N.: *Valence and the Structure of Atoms and Molecules*, Chemical Catalogue Company, Washington, 1923.

Linus, Pauling: *Linus Pauling in His Own Words*, Simon & Schuster, New York, 1995.

Mario, A. Morselli: *Amedeo Avogadro: A Scientific Biography*, Kluwer Academic Publishers, Boston, 1984.

Martha, Moore Trescott: *The Rise of the American Electrochemicals Industry, 1880–1910*, Greenwood Press, Westport, 1981.

Maurice, Crosland P.: *Gay-Lussac: Scientist and Bourgeois*, Cambridge University Press, New York, 1978.

Mead, Clifford and Hager Thomas: *Linus Pauling: Scientist and Peacemaker*, Oregon State University Press, Corvallis, 2001.

Michael, Hunter: *Robert Boyle Reconsidered*, Cambridge University Press, New York, 1994.

————: *Robert Boyle, by Himself and His Friends: With a Fragment of William Woton's Lost Life of Boyle*, William Pickering, Brookfield, 1994.

Mikael, Hard: *Machines Are Frozen Spirit: The Scientification of Refrigeration and Brewing in the 19th Century*, Westview Press, Boulder, 1994.

Mirkin, M.V. and Bard A.J.: *Scanning Electrochemical Microscopy*, CRC Press, Boca Raton, 1995.

Moore, Walter: *Schrodinger: Life and Thought*, Cambridge University Press, Cambridge, 1989.

Morris, William Travers: *A Life of Sir William Ramsay*, E. Arnold, London, 1956.

Mulliken, Robert S.: *Life of a Scientist: An Autobiographical Account of the Development of Molecular Orbital Theory*, Springer-Verlag, Berlin, 1989.

Mulliken, Robert S.: *Nobel Lectures: Chemistry, 1963-1970*, Amsterdam, Elsevier, 1972

Murray, Campbell and Harrison Hatton: *Pioneer in Creative Chemistry*, Appleton-Century-Crofts, New York, 1951.

Naomi, Pasachoff: *Marie Curie and the Science of Radioactivity*, Oxford University Press, New York, 1996.

Nye, Mary J. *From Chemical Philosophy to Theoretical Chemistry: Dynamics of Matter and Dynamics of Disciplines, 1800-1950*, University of California Press, Berkeley, 1993.

O. Bertrand Ramsay: *Van't Hoff–Le Bel Centennial*, American Chemical Society, Washington, 1975.

Otto, Hahn: *My Life: The Autobiography of a Scientist*, Herder and Herder, New York, 1970.

Paradowski, Robert: *The Structural Chemistry of Linus Pauling*, University of Michigan, 1972.

Paul, De Kruif: *Microbe Hunters*, Harcourt, Brace, New York, 1927.

Pauling, L. and Wilson E.B.: *Introduction to Quantum Mechanics with Applications to Chemistry*, Dover Publications.

Pauling, L.: *The Nature of the Chemical Bond*, Cornell University Press.

Pauling, Linus and Goudsmit Samuel: *The Structure of Line Spectra*, McGraw-Hill, New York, 1930.

Pauling, Linus and Hayward Roger: *The Architecture of Molecules*, W.H. Freeman and Co., San Francisco, 1964.

Pauling, Linus and Wilson Bright E.: *Introduction to Quantum Mechanics with Applications to Chemistry*, Dover Publications, New York, 1963.

Pauling, Linus: *General Chemistry*, W.H. Freeman and Co., San Francisco, 1947.

————————: *The Nature of the Chemical Bond and the Structure of Molecules and Crystals: An Introduction to Modern Structural Chemistry*, Cornell University Press. Ithaca. 1960.

Pearce, Williams L.: *Michael Faraday*, Da Capo, New York, 1965.

Ramsay, D.A. and Hinze J.: *Selected Papers of Robert S. Mulliken*, University of Chicago Press, Chicago, 1975.

Raymond, Szymanowitz: *Edward Goodrich Acheson*, Vantage Press, New York, 1971.

Robert, Boyle and Marie Boas Hall: *Robert Boyle on Natural Philosophy: An Essay with Selections from his Writings*, Greenwood Press, Westport, 1980.

Robert, Scott Root-Bernstein: *The Ionists: Founding Physical Chemistry, 1872–1890*, Princeton University, 1980.

Rosalind, Pflaum: *Grand Obsession: Madame Curie and Her World*, Doubleday, New York, 1989.

Rose-Mary, Sargent: *The Diffident Naturalist: Robert Boyle and the Philosophy of Experiment*, University of Chicago Press, Chicago, 1995.

Ruth, Lewin Sime: *Lise Meitner: A Life in Physics*, University of California Press, Berkeley, 1996.

Serafini, Anthony: *Linus Pauling: A Man and His Science*, Paragon House, New York, 1989.

Servos, John W.: *Physical Chemistry from Ostwald to Pauling*, Princeton University Press, Princeton, 1990.

——————: *Solid State and Molecular Theory: A Scientific Biography*, Wiley Interscience, New York, 1975.

Sheldon, Jerome Kopperl: *The Scientific Work of Theodore William Richards*, University of Wisconsin, 1970.

Sommerfeld, Arnold: *Atomic Structures and Spectral Lines*, Methuen, London, 1923.

Susan, Quinn: *Marie Curie: A Life*, Addison Wesley Longman, Reading, 1995.

Ted, Goertzel and Ben Goertzel: *Linus Pauling: A Life in Science and Politics*, Basic Books, New York, 1995.

Theodor, Benfey: *From Vital Force to Structural Formulas*, Chemical Heritage Foundation, Philadelphia, 1992.

——————: *Kekulé Centennial*, American Chemical Society, Washington, 1966.

Thomas, Hager: *Force of Nature: The Life of Linus Pauling*, Simon & Schuster, New York, 1995.

Thomson, G. P.: *J. J. Thomson, Discoverer of the Electron*, Anchor Books,. Garden City, 1966.

Truman, Schwartz and John McEvoy: *Motion toward Perfection: The Achievement of Joseph Priestley*, Skinner House Books, Boston, 1990.

Unwin, Patrick R. and Edwards Martin A. Martin S.: *Scanning Electrochemical Microscopy: Principles and Applications to Biophysical Systems*, Physiological Measurement, 2006.

Virginia, Veader Westervelt: *Incredible Man of Science*, Messner, New York, 1968.

Whitehead, Don: *The Dow Story: The History of the Dow Chemical Company*, McGraw-Hill, New York, 1968.

William, Brock H.: *Justus von Liebig: The Chemical Gatekeeper*, Cambridge University Press, New York, 1997.

Wotiz,. John: *The Kekule Riddle: A Challenge for Chemists and Psychologists*, Cache River Press, Vienna, 1993.

Zvonimir, B. M. and Eckert-Maksiæ, M.: *Molecules in Natural Science and Medicine: An Enconium for Linus Pauling*, Ellis Horwood, New York, 1991.

Zvonimir, B. M. and Orville-Thomas W.J.: *Pauling's Legacy: Modern Modeling of the Chemical Bond*, Elsevier, Amsterdam, 1999.

Index

D

E

K

M

N

O

T

V

W

X

Z

❑❑❑